About the Authors

Rudolf H. Moos, Ph.D., is Research Career Scientist and Director of the Center for Health Care Evaluation at the Palo Alto Department of Veterans Affairs Medical Center. He is also Professor of Psychiatry and Behavioral Sciences at the Stanford University School of Medicine.

Sonne Lemke, Ph.D., is Research Psychologist at the Center for Health Care Evaluation, Palo Alto Department of Veterans Affairs and Stanford University Medical Centers.

GROUP RESIDENCES FOR OLDER ADULTS

Group Residences For Older Adults

Physical Features, Policies, and Social Climate

RUDOLF H. MOOS
SONNE LEMKE

New York Oxford
OXFORD UNIVERSITY PRESS
1994

Oxford University Press

Oxford New York Toronto
Delhi Bombay Calcutta Madras Karachi
Kuala Lumpur Singapore Hong Kong Tokyo
Nairobi Dar es Salaam Cape Town
Melbourne Auckland Madrid

and associated companies in
Berlin Ibadan

Published by Oxford University Press, Inc.,
200 Madison Avenue, New York, New York 10016

Library of Congress Cataloging-in-Publication Data
Moos, Rudolf H., 1934–
Group residences for older adults :
physical features, policies, and social climate /
Rudolf M. Moos and Sonne Lemke.
p. cm.
Includes bibliographical references and index.
ISBN 0-19-506257-4
1. Old age homes—United States—Evaluation.
2. Life care communities—United States—Evaluation.
3. Nursing homes—United States—Evaluation.
4. Congregate housing—United Sttes—Evaluation.
I. Lemke, Sonne. II. Title.
HV1454.2.U6M66 1994 362.6'1—dc20 93-32701

9 8 7 6 5 4 3 2 1
Printed in the United States of America
on acid-free paper

To the memory of Ann Margulies,
whose ideas raised our work to a higher level

Preface

For as long as old age has been considered a separate research area, a major focus has been on factors that contribute to older people's quality of life. Among these are the individual's residential environment, both physical and social. The impact of the residential environment is of particular concern for the many older people who live in group settings. These settings provide a shared environment, serve a more vulnerable segment of the older population, result from intentional efforts on the part of architects, planners, and administrators, and often receive public funds.

In this book, we lay out a framework for understanding these residential environments and describe a program of research in which we obtained detailed information about more than 300 group settings from throughout the United States. The first step in this research program was to develop measures for characterizing important aspects of these settings; the result was the Multiphasic Environmental Assessment Procedure (MEAP). We use descriptive information about the physical resources, policies and services, and social climate of facilities to help us understand how different aspects of the environment are related, how resources are allocated to older people based on their social characteristics and curent functioning, how environmental factors influence the lives of residents, and how these residential settings might be improved.

We are indebted to Paul Haber, whose vision helped stimulate the initial research proposal. He saw that techniques for systematically characterizing residential and treatment environments, which had been developed in earlier research in dormitories and psychiatric programs, could be adapted and have useful applications in residential programs serving older veterans.

This research program was supported in part by Department of Veterans Affairs (VA) Medical and Health Services Research and Development Service (HSR&D) research funds. Shirley Meehan (then Garin) represented the VA HSR&D Service in an arduous and searching site visit on the initial proposal. This eventually led to the initiation of the project, continued renewal funding, and the development of a long and fruitful association with Shirley and the VA HSR&D Service.

The work was also supported by NIMH grant MH28177 and NIAAA grant AA06699. Overall, the funding allowed us to develop measurement instruments, test their use in a large sample of facilities, revise them on the basis of these data, and then apply the revised instruments in a nationwide set of facilities. It also allowed us time to consider the meaning of these data and how best to present our findings to others. We gratefully acknowledge this financial support.

A project of this temporal and geographic scope represents the efforts of many individuals and organizations, and we welcome this opportunity to thank them for their contributions. First, we wish to thank the residents and staff in the participating facilities. Although the project offered few direct benefits to them, they gave generously of their time and effort.

Over the years, a number of research assistants contributed their effort and skills to the project. Mary Gauvain and Barbara Mehren helped develop the initial instruments and, along with Wendy Max, collected data in the California sample and worked on revision of the MEAP.

Jane Clayton, Joan Kahn, and Eric Postle helped collect data to address issues of interobserver reliability and test–retest stability and carried out detailed observations in a facility that was relocating to a new building, discussed in chapter 10. Margaret Stevens and Margareta Halstead, older women living in the local community, also completed facility assessments to help address issues of interobserver reliability.

Jane Clayton and Diane Denzler helped coordinate data collection in the national sample and in veterans facilities from throughout the United States. In this effort, they were assisted by many individuals who served as local contacts and data collectors.

We would like to acknowledge specifically the contributions of Erin Gardner, Melodie Harwell Page, and Joanie Weldon, who worked under the direction of Eugene Waters (Georgia); April Crusco, who worked with Ed Trickett (Maryland); Patricia LeFrapper, who worked with Tom Byerts (Illinois); and Harold Braithwaite and Charles Wadsworth, who worked with Abe Wandersman (South Carolina). Juva DuBoise (Oregon), Dennis Geehan (Ohio), Terence Joiner (Missouri), Judith Jurmann (New York), Ruth Lederman (Pennsylvania), Virginia Moomey (Nebraska), and Ivy Snowden (South Dakota) also completed facility assessments in their local areas.

We also benefited from the willingness of other researchers to share data from research projects using the MEAP. These include Lee Hyer (North Carolina), Patricia Parmelee (Utah), staff of the Adult Congregate

Living Facility Project at the Florida Mental Health Institute, and the Quality of Life Task Force of the California Association of Homes for the Aging.

Margaret W. Linn and Lee Gurel used the MEAP in nursing homes and VA units as part of their VA Cooperative Study No. 111 "Mental Patients and Nursing Home Care." They kindly shared these facility data. Local data collectors included Bernadette McDonnell and Eleanor Tarpy (Brockton, Massachusetts), Leah Hostetter and Joyce Vaughan (Coatesville, Pennsylvania), Cynthia Snodgrass (Ft Lyon, Colorado), Susan Smith and Linda Vernon (Hampton, Virginia), Bernard Schut and Jodi Wahl (Knoxville, Iowa), Rosalyn Sprague, Debbie Vance, and Vicki Woods (Marion, Indiana), Rainelle Holcomb and Deborah Wagner (Perry Point, Maryland), and Ruth Watkins and Shelly Wolff (St. Cloud, Minnesota).

Tom David, a postdoctoral fellow with out research group, helped collect data on interobserver reliability and stability over time and on the impact of a facility's relocation. In addition, he and Diane Denzler organized the collection of data on preferences for features in congregate settings. Carolyn Lamb joined us for a summer and collected preference data in additional nursing home and residential care facilities.

Throughout the project's existence, many people contributed to data management and analysis. In particular, we wish to thank Bernice Moos, who has been a guiding force from the beginning and who conducted the final analyses reported in this book. And we thank others who fulfilled similar roles along the way, beginning with Wendy Max and extending through Joan Kahn, Diane Denzler, Eric Postle, Sarah Buxton, Amy Marder, Ann Marten, Jon Hussey, and Philip Granof.

Penny Brennan, Tom David, Amnon Igra, and Christine Timko coauthored earlier papers based on some of these data; their formulations and approaches influenced the way in which we present our findings here.

Early versions of this book were in the able hands of Adrienne Juliano, whose speedy word processing and attention to detail were much appreciated. Debbie Davis was responsible for word processing later revisions of the manuscript; we appreciate particularly her skills with figures and tables.

A number of people contributed critical comments as the book progressed toward its final form, including Penny Brennan, Virginia Junk, Marilyn Skaff, and Christine Timko. Ann Margulies provided detailed editorial suggestions that helped improve the clarity of specific aspects of the manuscript and the organization and consistency of the book as a whole.

We hope that our efforts to assess residential settings and understand their impacts will ultimately lead to improvements in older people's lives.

Palo Alto, California

R.H.M.
S.L.

Contents

I

CONCEPTUAL OVERVIEW AND DESCRIPTION OF SAMPLE

1

A Conceptual Framework for Evaluation

About 10% or 3 million of the 30 million people in the United States who are over age 65 live in a group residential setting. Of these 3 million older adults, almost half live in nursing homes; the rest live in varied types of residential care, community care, and supportive apartment facilities. However, a much higher percentage of older people will live in such settings at some point in their lives. According to one estimate (Kemper & Murtaugh, 1991), an older person faces a 43% lifetime risk of at least temporary placement in a nursing home and a risk greater than 20% of spending a year or more in a nursing home. Demographic and social trends suggest that residential programs will grow in importance during the next few decades (Brody, 1982; Kane & Kane, 1987) and will touch the lives of increasing numbers of people.

These social trends have been paralleled by a body of research on group housing and its impact on older persons. Some of this research focuses on taxonomic and descriptive issues, such as defining the types and characteristics of group housing. Other research considers the relationship between group housing and resident characteristics and outcomes. What conditions contribute to the older person's use of such housing? What effects do these settings have on the people who live in them? Who benefits from a particular housing program? We have addressed these issues in an integrated program of research on congregate residential facilities.

In order to structure this presentation, we first describe the types of group housing and provide a brief history of the development of group residential settings for older people. We then present the conceptual framework that has guided our work. This framework highlights the need to conceptualize and measure program characteristics and to consider different types of residential programs within a common model. In addition, the framework focuses on the interrelationships of program characteristics and on the match between individual resident characteristics and the program environment. Our work is based on the premise that congregate residential settings can make an important difference in the quality of older people's lives and that information about the design and impact of group residences can help managers and practitioners improve these settings.

TYPES OF CONGREGATE LIVING ENVIRONMENTS

We use a number of terms interchangeably to delimit the range of settings where older people reside with nonrelatives and thereby have access to services. These terms include *group* or *congregate residences, residential programs,* and *sheltered care settings.* At one end of this continuum, we include hospitals for the chronically ill, skilled and intermediate care nursing facilities, and other long-term care settings. At the other end are settings that provide minimal services, such as a meal program and weekly housecleaning. We exclude from this category settings where stays are short-term (acute care hospitals and hotels), where residents are related (single-family homes), or where no services beyond shelter are provided (most apartments and single-room-occupancy hotels, mobile-home parks, and some retirement communities).

We group these disparate forms of housing together because we believe that they can be usefully compared on a number of common dimensions. By bringing unrelated individuals together in a residential setting and giving them access to services, these settings create a stage for distinctive social processes. Individuals in these settings must deal with an additional layer of social organization. As Markson (1982, p. 52) puts it, "While monasteries obviously differ from college campuses, and a residential congregate-complex for those sixty-five and over is distinctly different from a skilled-nursing facility, all have more common facilities, rules and regulations, social norms, and encourage closer contact with other residents than does traditional community living."

For specific purposes, however, distinguishing between subgroups of residential programs is useful, and we have opted to do so in terms of three levels of care: nursing homes, residential care facilities, and congregate apartments. For our purposes, nursing homes are settings in which a majority of residents have regular access to professional nursing services and in which housekeeping, meals, and personal care are routinely provided. Although residential care facilities also provide routine housekeeping, meals, supervision, and assistance with activities of daily living for the majority of residents, these facilities offer only limited health services. Congregate apartments typically provide at least a meal program but do not give routine care to the majority of residents.

HISTORICAL OVERVIEW OF RESIDENTIAL PROGRAMS

In preindustrial Western societies, few individuals lived into old age. Among those who survived infectious disease and the hard conditions of life, the sick or poor were expected to rely on their families; where the family failed in its duty, these older people turned to the church or one of the handful of benevolent associations that provided care for people in need.

Centralization and Decentralization of Residential Facilities

During the Industrial Revolution, governments began assuming limited responsibility for elderly people but did so grudgingly. In England, for example, the Elizabethan poor laws gave local governments responsibility for administering almshouses, but conditions in these almshouses were deliberately unpleasant in order to discourage all but the most destitute from using them.

In the American Colonies, which followed the example of the Elizabethan poor laws, the most prevalent form of government assistance was payment to the family or to people providing housing for nonrelatives. In the decades following the Revolutionary War, local governments began to establish almshouses in order to centralize and thereby reduce the cost of care. In addition to saving money by their economy of scale, these institutions were designed to be financially self-sufficient by relying on the residents' labor (Sherwood & Mor, 1980).

During the second half of the nineteenth century, the model of specialized housing for the older person was the large, centralized institution. As the century progressed, local governments were more willing to provide for specific groups of the elderly, including widows and soldiers. For example, after the Civil War, Congress established the National Asylum for Disabled Volunteer Soldiers, which evolved into the National Homes and later into the Department of Veterans Affairs (VA) domiciliaries. States, too, began to establish homes for veterans unable to maintain themselves in the community. Private philanthropy also grew during the nineteenth century. By the 1920s, nonprofit homes for the aged had become the common model of care for the elderly; these facilities were supplemented by psychiatric hospitals and county homes.

A growing distrust of large institutions manifested itself in the Social Security Act of 1935. Under this act, the federal government made payments directly to indigent older persons to purchase housing and other services; public institutions were barred from receiving these monies directly. During the 1930s and 1940s, increasing numbers of older people became boarders in private homes or entered small, family-run rest homes or sanitariums. As these residents aged and became more disabled, nursing care was often added to the services. In part because of the availability of public money, the number of such facilities increased markedly. In the 1950s, the Social Security Act was amended to allow federal assistance for care in institutions and to require state agencies to supervise all medical settings receiving federal funds. (For more detailed histories, see Achenbaum, 1978; Brody, 1977; Vladeck, 1980.)

The Development of Modern Nursing Homes and Residential Care Facilities

A striking change since the 1950s is the growth in both the number and size of nursing homes. This growth has been traced to the aging of the population, the introduction of programs of federal support and their accompanying regulations, and the movement to deinstitutionalize mental hospital patients (Dunlop, 1979). Nursing homes have also changed qualitatively as nursing has achieved stature as a health profession and as federal and state regulations have been imposed. At the time of the 1985 National Nursing Home Survey, about 1.3 million persons over age 65 were living in facilities that provided professional nursing and personal care services (Hing, 1987).

Although nursing homes and residential care facilities evolved from common sources (private sanitariums and old age homes), they now comprise distinctive subgroups. With the advent of more direct financial assistance to the elderly poor, the number of well but indigent people in institutions has decreased. In addition, more government assistance is available to help older people with physical impairments to maintain their individual households and to place older people with severe impairments in settings with professional nursing care. Together, these factors help account for the relatively stable numbers in residential care facilities over recent decades.

The impact of these policies can be illustrated with figures on the populations of VA domiciliaries and nursing home units. In 1953, about 17,000 veterans lived in VA domiciliaries. By 1966, the figure had dropped to 14,000, and by 1980, it was down to about 8,000. The number is now increasing again as the domiciliaries serve the growing number of World War II veterans who are 65 and over. In contrast, VA nursing home units have served ever-increasing numbers: 4,500 average daily census in 1971, 7,900 in 1980, and more than 13,000 in 1990 (Department of Veterans Affairs, 1991; Mather, 1984: Veterans Administration, 1981).

Regulation of residential care facilities, unlike nursing homes, is not uniform and is altogether absent in some states. As a consequence, residential care facilities are more varied, and statistics on their use are more difficult to obtain. In a national survey, Mor, Sherwood, and Gutkin (1986) identified 118 government programs that regulate more than 29,000 residential facilities whose residents include elderly persons; these facilities have more than 370,000 residents (see also Gottesman, Peskin, & Kennedy, 1990).

Modern Congregate Housing for Functionally Independent Older Adults

First authorized in 1956, federal programs of housing assistance for functionally independent older adults now serve about 4% of the older population. These residences provide varied levels of services; some are indistinguishable from independent apartments except for age concentration, whereas others are associated with programs offering extensive support services. Similar housing programs have also been developed without federal assistance.

Thus, different subgroups of facilities have evolved in response to the needs of older adults who are frail, indigent, or otherwise impaired in their ability to maintain an independent household.

CONCEPTUAL PERSPECTIVES

Three assumptions guide our approach to understanding congregate living environments and their impacts. First, in order to examine the influence of residential settings on health and well-being, *we need new ways to characterize the salient aspects of residential programs.* Although most gerontologists endorse the idea that both personal and environmental factors determine behavior, evaluation research has typically conceptualized the residential program as a "black box" intervening between resident inputs and outcomes. For example, the housing program is often assessed only in terms of broad categories, such as the level of care provided or whether the setting is age-segregated. To enrich descriptions of these settings, we propose some new ways to measure program characteristics; these measures enable us to identify specific aspects of residential programs and to analyze their influence on residents' activity levels, well-being, and use of program services.

Our second assumption is that although group residences for older people are diverse, *a common conceptual framework can be used to evaluate residential programs, and doing so has several advantages.* The framework allows us, for example, to identify similar processes occurring in different types of settings and to specify the extent of environmental change an individual experiences when moving from one type of setting to another.

Our third assumption is that *more emphasis should be placed on the process of matching personal and program factors and on the connections between person-environment congruence and resident outcomes.* To understand the influence of group living settings more fully, we need to examine the processes of self-selection and social allocation that affect how older people are matched to residential environments. We also need to focus on how facility policies and services vary in their impact on residents of different ability levels. Although researchers have recognized the complexity of person–environment transactions, empirical work has not adequately reflected the multicausal, interrelated nature of the processes involved.

An Integrative Conceptual Framework

The model shown in Figure 1.1 follows these guidelines and provides a conceptual framework for examining group residential settings and their influence on older people. In this model, the connection between the objective characteristics of the residential program (panel I), personal factors (panel II), and resident adaptation (panel V) is mediated by the program social climate (panel III) and by residents' coping responses (panel IV). The model specifies the domains of variables that should be included in a comprehensive evaluation. The model can also guide more focused evaluations and serve as a framework for integrating research results.

The objective characteristics of the program (panel I) include the aggregate characteristics of the residents and staff, the physical design of a

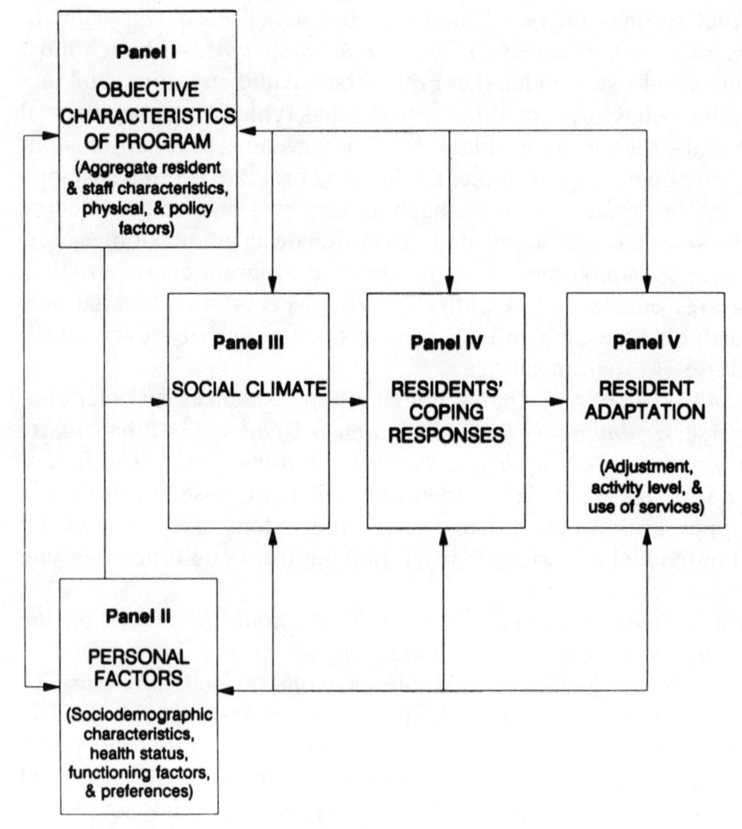

Figure 1.1 A model of the relationship between program and personal factors and resident stability and change.

program, and the program's policies and services. These three sets of objective environmental factors combine to influence the quality of the program culture or social climate (panel III). The social climate is part of the environmental system, but we place it in a separate panel to highlight its special status: The social climate is in part an outgrowth of the objective environmental factors and also mediates their impact on resident functioning.

Personal factors (panel II) encompass an individual's sociodemographic characteristics and such personal resources as health, cognitive status, and functional ability. They also include an individual's preferences for specific characteristics of residential facilities. The environmental and personal systems influence each other through selection and allocation processes. For example, most facilities select new residents on the basis of personal and functional ability criteria. Similarly, most older people have some choice about the facility they enter. In general, these selection processes tend to enhance the congruence between personal and facility characteristics.

Both personal and environmental factors affect an individual's coping responses (panel IV). In turn, coping responses influence such outcome indices as the resident's adjustment, activity level, and use of program services (panel V). For example, special activity areas and organized social activity programs (panel I) may contribute to a cohesive social climate (panel III). In such a setting, a new resident may be more likely to seek information about facility programs and to join a resident activity group (panel IV) and, ultimately, to show higher morale (panel V). Personal characteristics such as age, sex, and functional ability (panel II) can influence an individual's view of program cohesion (panel III), propensity to participate in an activity program (panel IV), and morale and adjustment (panel V).

The model shows that resident adaptation is also affected directly by stable personal factors. For example, residents who are in better health when they enter a facility are likely to be in better health a year later. Residential programs may have some direct effects as well, as when an individual experiences better health because of the good medical care provided in a setting.

Finally, the model depicts the ongoing interplay between individuals and their residential environment. Residents who make more social contacts or participate in planning activities may experience improved self-confidence (a change in the personal system). Residents who voice their preference for more choice in their daily activities may help initiate more flexible policies (a change in the environmental system). More generally, individual outcomes reenter the model in the form of aspects of the personal system. Individual outcomes also contribute to defining the environmental system; for example, when the health outcomes for all individuals in a setting are considered together, they constitute one aspect of the suprapersonal environment.

Person–Environment Congruence Models

Our framework notes that people and programs influence each other. In addition, a given environmental feature can vary in the way it affects different residents. High expectations for independent behavior may promote better functioning among more competent residents but may have little influence or even result in lower morale among impaired residents. In contrast, providing supportive services may benefit impaired residents but may have little or no impact on more competent residents. Other theorists have developed two general models to account for the differential influence of program characteristics on health and well-being among individuals who vary in their personal competencies and needs; we describe these models, then build on them to formulate the integrated approach we use here.

Environmental Demands and Personal Competencies

Lawton's (1982, 1989) perspective, which focuses on three panels in our model, relates environmental demands (panel I) and personal competence (panel II) to emotional responses and adaptive behavior (panel V). Personal competence is a selected array of characteristics that defines the individual's functional capacity, including physical health, mobility status, and cognitive ability. Lawton characterizes the environment in terms of its objective demands or its potential for activating behavior. An environmental condition is said to have a given demand quality if it is associated with a particular behavioral outcome for any group of individuals. For example, if physical barriers to access reduce activity levels among impaired elderly residents, then such barriers can be seen as environmental demands for all residents. If proximity to age peers increases social interaction, then proximity is considered an environmental demand.

Lawton's model assumes that each individual has a level or zone of environmental demand that permits maximum comfort and performance. In general, demands that match an individual's abilities relatively well tend to elicit positive affect and adaptive behavior; demands that are either too weak or too strong tend to elicit negative affect and less effective behavior.

Lawton and his colleagues have developed several predictions from this model. For example, the environmental docility hypothesis proposes that variations in environmental demands influence the behavior of less competent individuals more strongly than that of more competent individuals. More specifically, people with fewer personal resources are more vulnerable to variations in environmental demands because they can achieve their optimum functioning in only a relatively narrow range of settings. In contrast, people with intact abilities can function effectively in a wider range of environments.

One controversial element of Lawton's model is the decline in functioning predicted to occur when the environment is too rich in resources or places too few demands on the individual. For example, Lawton's model predicts that ambulatory adults will show a decline in functioning when the environment is too accessible. The logic is that people need a specific level of environmental challenge to stay engaged with the environment and to exercise their functional capacities. Thus, ambulatory residents who use elevators (an environmental resource) rather than stairs may show a premature decline in functional ability. Similarly, when residents who can walk with assistance are transported in wheelchairs, they are likely to show a decline in strength and coordination.

Carp and Carp (1984) have also focused on the concepts of demand and competence, which they see as relevant primarily to life maintenance needs. Environmental resources, such as prosthetic features (panel I), and individual competencies (panel II) influence how well an individual can perform the daily activities required to satisfy life maintenance needs (panel V). More competent people are better able to satisfy their life maintenance needs, and it is easier for people to satisfy life maintenance needs in environments with more resources (fewer barriers to behavior). In addition to the direct impact of environmental resources and individual competence, person–environment congruence may account for additional variance in behavior and need satisfaction.

In this conceptualization, person–environment congruence is the degree of complementarity between a person's competence and relevant environmental resources. People who have a low level of competence (such as residents who need assistance in walking) may need a prosthetic environment (physical features to aid mobility) to achieve their maximum level of adaptation. The Carps' model does not incorporate the concept of functional decline in an environment with too many resources.

Environmental Resources and Personal Needs

The competence–demand model, which tends to treat competence and demands as global dimensions, has not explicitly paired specific competencies with specific demand qualities of the environment. More important, as Carp (1978–79) noted, the model is limited because it focuses on personal competencies to respond to demands rather than personal preferences for opportunities. To address this issue, some theorists have formulated models that consider the congruence between environmental resources and personal needs.

Kahana (1982) identified seven areas in which environmental resources may be congruent or incongruent with personal needs. Three are based on the setting characteristics formulated by Kleemeier (1961) and reflect the level of autonomy permitted in an environment, the homogeneity and continuity of the environment, and the extent to which staff encourage residents' dependency. Correspondingly, individuals vary in their need for autonomy, continuity, and a well-structured environment. The four other dimensions reflect personal factors that tend to show characteristic changes with age: need for activity, open expression of affect, tolerance of ambiguity, and impulse control. The corresponding environmental factors are opportunities for involvement, encouragement of emotional expression, clarity of structure and expectations, and emphasis on impulse control (see also Harel, 1981).

Carp and Carp (1984) have combined the competence–demand and needs–resources approaches into a two-part model of the determinants of well-being. As noted earlier, the first part of the model considers competence and resources in relation to life maintenance needs. This part of the model is based on the belief that individual functional capacities influence the outcome of a limited range of person–environment transactions among older people who are below some competence threshold. However, for older people who can manage their environment, differences in functional ability are less important in determining outcomes than are higher-order needs and personal preferences.

The second part of the model focuses on these higher-order needs and aspects of the environment that affect their satisfaction. Examples of higher-order needs are needs for autonomy, social affiliation, and emotional expression. According to Carp and Carp (1984), such personal needs differ from competencies in that they are not directly related to better adaptation; that is, people with a higher need for autonomy or for affiliation do not necessarily experience better well-being. Moreover, environmental resources that satisfy such needs are not necessarily related to better adaptation. In other words, the model is neutral about the value of these personal needs and environmental characteristics. Neither personal nor environmental factors by themselves account for much of the variance in adaptation; instead, the major issue is the level of person–environment congruence.

Partially in response to these ideas, Lawton (1989) has recently expanded his model to focus explicitly on the interaction between personal and environmental resources and environmental proactivity. This formulation views the individual as active in seeking opportunities for personal satisfaction. Individuals with more personal resources are able to make use of a broader range of environmental resources; the environment is richer in opportunities because the more competent individual has the personal resources to make use of them.

An Integrative Model of Environmental and Personal Characteristics

Our approach integrates these earlier models and, in doing so, purposely forgoes some specificity. For example, we believe that environmental demands and resources are interconnected, as are personal competencies and needs. These apparently diverse factors can be usefully considered within one conceptual framework.

We opt to describe environmental conditions primarily as resources or opportunities; however, there is no clear a-priori way to distinguish resources from demands. Most environmental conditions can be both resources and demands, depending on the context in which they occur, the functional abilities and personal inclinations of the individuals who appraise and cope with them, and the outcomes that are examined. Organized activities or policies designed to enhance personal autonomy can be conceptualized as resources, but they also have a demand quality. The presence of a physical amenity, such as a comfortable lounge, is a resource; it may also create a demand for activity and social interaction. A cohesive social climate can be a supportive resource, but it may also exert a press for involvement and conformity.

In focusing on environmental resources, we assume that environmental conditions relevant to higher-order needs, such as policies that promote autonomy and affiliation, generally are directly associated with better adaptation. Thus, facilities that have more physical amenities and social–recreational aids are likely to enhance activity levels and well-being for most residents. We make the same prediction for facility policies that promote residents' personal control. However, as noted earlier, person–environment congruence is also important in that some functionally impaired residents may prefer a structured setting and may find it hard to maintain well-being in a facility that provides a high level of personal choice.

With respect to personal factors, impaired individuals often reappraise their needs to bring them in line with a decline in their abilities. Thus, personal competencies, needs, and preferences are interrelated; for example, the need for autonomy or affiliation may decline as physical or mental impairment increases. Moreover, in contrast with Carp and Carp's (1984) assertion, we believe that personal characteristics (needs), such as high extroversion and independence, are similar to functional competencies in that they are associated with good adaptation in most environments.

Finally, we think that environmental conditions should be conceptualized independently of the adaptational outcomes that may be associated with them. Our approach emphasizes environmental conditions that are likely to have important implications for adaptation. However, we first conceptualize and measure these conditions, then focus on their empirical associations with varied outcomes.

Prosthetic Features and Residents' Mobility Status

Figure 1.2 illustrates these ideas by depicting our predictions of how physical features that aid mobility can influence activity levels for residents who vary in mobility status. As a group, ambulatory residents are active irrespective of the level of physical features to aid mobility; that is, the absence of such features does not reduce their activity levels. Among residents who use a wheelchair, however, supportive physical features are

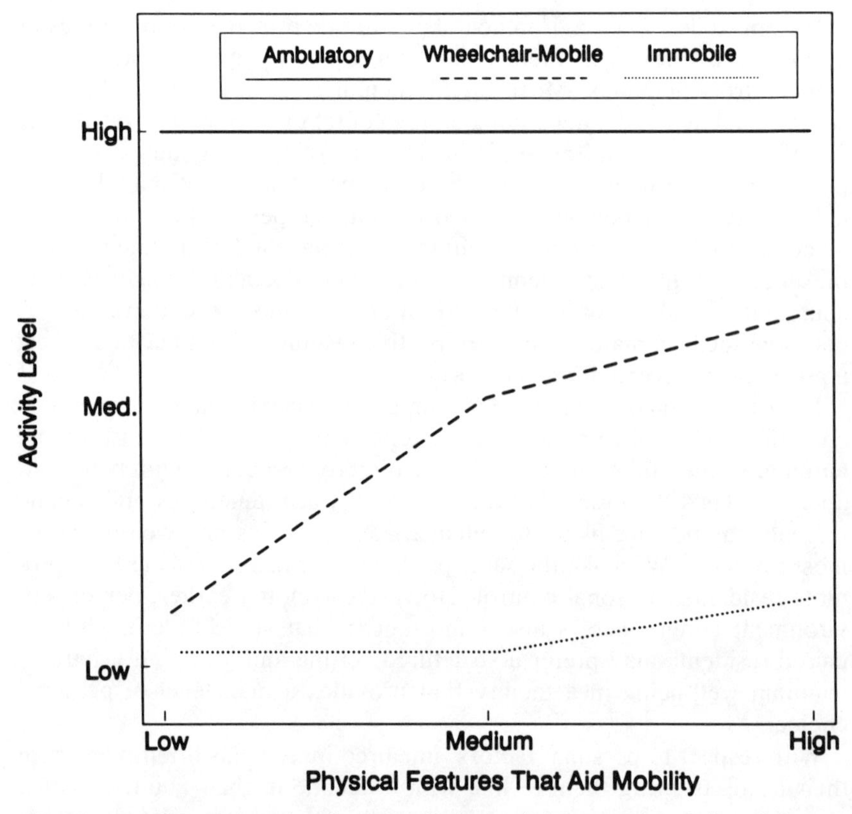

Figure 1.2 The predicted influence of physical features that aid mobility on activity level.

associated with an increase in activity level. For immobile individuals, only major restructuring of the physical environment, such as provision of electronic controls and motorized wheelchairs, is likely to have an impact on activity level and then only raising it slightly; in addition, such residents need program interventions, such as staff assistance to get them out of bed, in order to benefit from a more accessible physical environment.

Figure 1.2 also shows that a change in environmental conditions (such as from low to medium physical resources to aid mobility) is likely to have a greater influence on the behavior of residents who use wheelchairs than on ambulatory or immobile people. Thus, a small improvement in a program can have a substantial positive impact on residents with moderate impairment because the environmental barriers place a cap on the expression of individual differences among them.

One important issue is whether ambulatory residents tend to show a reduction in activity level in a high-resource setting, as Lawton (1989) would predict. We believe that functionally competent people will seek needed levels of stimulation and that such decrements are likely to occur only when such efforts are systematically blocked by the environment.

Policies That Support Personal Control and Residents' Need for Autonomy

Figure 1.3 illustrates our predictions of how policies that support personal choice and control can influence residents who vary in functional ability. Functionally able residents experience better adjustment in facilities with policies that support personal choice and control. Residents with some functional impairment also adjust better in settings that provide more personal control, but the level of personal control is not as closely related to their adjustment as it is among functionally more competent individuals. Facility policies have relatively little effect on adjustment among residents who are quite impaired. One question here is whether very impaired

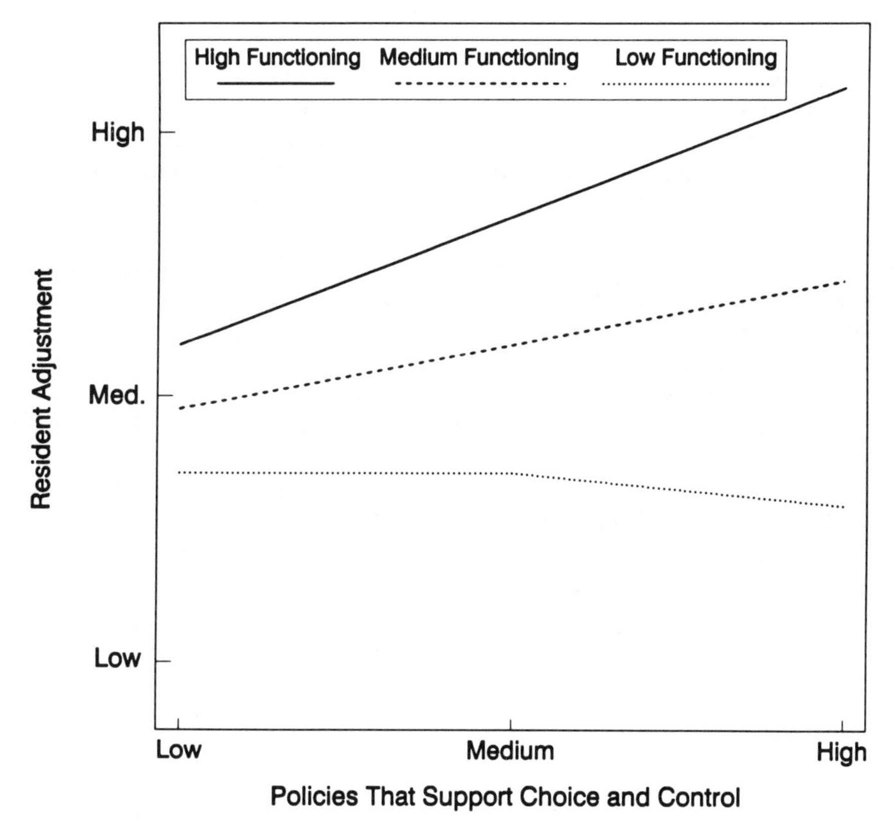

Figure 1.3 The predicted influence of policies that support choice and control on resident adjustment.

residents experience poorer adjustment in autonomy-enhancing environments, as Lawton (1989) would predict. As shown in Figure 1.3, we believe that such a decrement in adjustment may occur but that it is likely to be relatively small because these environmental features provide opportunities that an individual can ignore if they are not relevant to his or her needs and abilities.

Although the predictions shown in Figures 1.2 and 1.3 differ somewhat in their specifics, they can be integrated within one general model. In both examples, resident outcomes (activity level and adjustment) are a function of personal factors, environmental factors, and person–environment congruence. In Figure 1.2, the more mobile residents show higher activity levels across all environmental conditions than do less mobile residents. Also, aggregate activity levels are higher in high-resource than in low-resource settings. Thus, personal and environmental factors each account for considerable variation in activity level. Person–environment congruence is also important in that variations in physical features affect some residents (those who need assistance in walking or who are in wheelchairs) more than others (ambulatory residents or immobile residents).

Similarly, in the example shown in Figure 1.3, residents who are functionally able show better adjustment across settings. Also, resident adjustment tends to be better in high-resource settings. Thus, personal and environmental factors alone account for a sizable amount of the variation in adjustment. In addition, person–environment congruence is important because variations in facility policies have different effects on residents who differ in their functional abilities. The set of individuals likely to benefit most from an intervention will vary from case to case. The proportion of variance in an outcome criterion accounted for by personal factors, environmental factors, and person–environment interactions will also vary depending on the people, settings, and outcomes studied. We believe that these variations can be accommodated within one general person–environment congruence model.

At present, we do not have a common metric for measuring persons and environments. Thus, we cannot specify the presence of person–environment congruence apart from either the subjective appraisal of the individual or the behavioral and affective outcomes of specific combinations of environmental and individual factors. We do not assume that subjective appraisal of person–environment congruence is synonymous with better outcome; we see this as an empirical issue. For example, perceived congruence may be associated with better outcomes in some areas (such as a general sense of satisfaction) but worse outcomes in others (such as less striving for independence). In general, we focus on the associations between specific combinations of environmental and personal factors and specific adaptational outcomes. We return to these issues in chapters 2, 8, 9, and 10.

Applications of the Model

These efforts to specify the relationship between person–environment congruence and outcomes illustrate how our general conceptual model can be elaborated in terms of specific hypotheses. The model can also help to organize the approach to a number of traditional issues in gerontology. For example, early work on the impact of institutionalization tended to focus on the personal attributes that characterize older people in some group residential settings. These attributes, such as apathy, withdrawal, and dependency, were treated as outcomes (panel V) and traced to the features of these settings (panel I) that distinguish them from independent living. In their review of this research, Kasl and Rosenfield (1980) pointed out that such studies are inconclusive because they do not take into account possible selection bias or the effects of events preceding institutionalization.

Similarly, early work on relocation focused on the stressful event of moving from one setting to another and on the subsequent functioning of the individual. Only later did researchers begin to examine the extent of change in the objective environmental system (panel I) and in social climate (panel III) that the resident experienced during relocation, or the personal characteristics (panel II) that predisposed the individual to particular outcomes (Lieberman & Tobin, 1983).

Research on quality of care also illustrates applications of the model. Quality of care has been defined in terms of structural, process, and outcome measures. In studies using structural or process measures, specific aspects of the program are taken to indicate good or poor quality of care. Thus, physical therapy might be defined as a beneficial element in a nursing home environment. Other indices of the structure of care (panel I) include measures of staffing levels, the presence of in-service programs, and various physical features of the facility. Process indices focus primarily on how policies are made, how social activities are planned and executed, and how the daily routine of life is organized (panel I), although sometimes aspects of the social climate (panel III) are also assessed. In contrast, outcome measures (panel V) focus on the individual and include the mood, behavior, functioning, and satisfaction of the consumers of care.

A growing number of studies consider the link between structural or process measures of quality of care (panels I and III) and outcomes (panel V). For example, Linn, Gurel, and Linn (1977) noted that better physical features, more nursing hours per resident, and better meal services were associated with better health outcomes. Wieland and his colleagues (Wieland et al., 1986) found that the improved quality and continuity of care following the reorganization of a nursing home along an academic nursing home model led to significant improvements in residents' functional status and morale. A more comprehensive examination of quality of care would also consider the individual (panel II) by examining who benefits from what aspects of care.

THE RESIDENTIAL PROGRAM

Knowledge about the residential program serves as a foundation for examining the interrelationships depicted in the model. Yet, aside from some valuable naturalistic descriptions of specific facilities (for examples, see Hochschild, 1973; Ross, 1977; Schmidt, 1990; Tisdale, 1987), few researchers have focused on the task of systematically assessing the residential program. One factor contributing to this neglect is the difficulty of selecting a measurement strategy. For example, a person's milieu can be described by cataloging specific features, by focusing on the arrangement or structure of these elements, or by specifying their function. Descriptive terms can range from those that are relatively close to the raw data to higher-order variables. Staff members, residents, or outside observers can report on the environment. Finally, measurement can focus on one or several distinct domains of environmental variables, ranging from physical features to social relations. We have identified four research traditions, each focusing on a different domain.

One tradition involves the use of aggregate personal characteristics as measures of environmental factors. This approach is based on the belief that the aggregate of the members' attributes (the suprapersonal environment) helps to define the subculture that develops in a group. This subculture is thought to influence the behavior of individual members. For example, Cohen, Bearison, and Muller (1987) suggest that the limited cognitive challenge and lack of diversity of social interaction in age-segregated housing may lead to a decline in interpersonal understanding and in the ability to empathize with another person's perspective.

As part of a second research tradition, architects and designers have considered such factors as the quality of the physical milieu, safety, accessibility to the handicapped, and site selection for security and community involvement (Lawton, 1975; Regnier & Pynoos, 1987). Some designers have offered guidelines for designing and furnishing settings for the elderly, particularly for those with disabilities (Koncelik, 1976, 1987; Steinfeld, 1987).

In keeping with a third research tradition, some sociologists and social psychologists have focused on the policies and services of group living settings. Some researchers have considered such indices as size and staffing levels, whereas others have examined complex organizational characteristics. For instance, Goffman (1961) described total institutions as those that combine ordinarily separate spheres of an individual's life (such as places of work, residence, and recreation), require residents to organize their life in terms of a fixed schedule, and limit social exchange with the outside world. Similarly, Kleemeier (1961) characterized specialized housing in terms of the enforced closeness of individuals and absence of privacy and the extent to which residents are subject to social control through restrictive rules and policies.

The social climate perspective is a fourth research tradition. In this view, discrete events are manifestations of underlying characteristics of the setting and its physical features and policies. For example, in a program that emphasizes resident influence, older people are likely to have a say in making the rules, suggestions are more likely to be acted on, and new ideas are likely to be tried out. We have identified three sets of social climate dimensions. Relationship dimensions assess the quality of interpersonal relations. Personal growth or goal orientation dimensions measure the directions of personal development emphasized in the setting. System maintenance and change dimensions deal with the degree to which the environment is orderly, clear in expectations, and responsive to change (Moos, 1987).

Measuring the Environmental Domains

Considering all these matters, a researcher who wishes to assess the residential environment is faced with numerous choices and an absence of clarifying conceptual models. Early environmental measures tended to use rating systems to focus on either specific aspects or the overall quality of care in a setting. For example, Bennett and Nahemow (1965) used Goffman's concept of totality to formulate a 10-item index that covers information such as the disposition of personal property, provisions for dissemination of normative information, and type of sanction system. Linn (1966) designed the Nursing Home Rating Scale to incorporate quality criteria used in licensing and accreditation and to differentiate among nursing homes in the areas of patient care, administration, staffing, and physical features.

The Quality Evaluation System (QES) represents a more systematic attempt to assess long-term care settings by focusing on facility resources and resident needs (Dennis, Burke, & Garber, 1977). The QES involves inspection of a facility by two members of an assessment team, interviews with two staff members, a questionnaire completed by the facility administrator, interviews with a sample of residents, and a review of their medical records. The QES measures three components of facility resources: institutional services, professional services, and psychosocial environment. Data on residents' needs for nursing, personal, and psychosocial care are obtained for a sample of residents and can be summarized as profiles of individual residents or aggregated into a profile of the facility's residents. This information can be used to audit the quality of care (Hewitt et al., 1985) and to determine how well a facility meets its residents' needs.

The Quality Assessment Index (QAI) is another method for measuring the quality of long-term care facilities (Gustafson et al., 1990). The QAI assesses seven aspects of structural, process, and outcome indices of quality, including the resident care process (medical records, resident influence, staff communication, and staff attitudes toward residents), staff resources, facility characteristics (floor plan, cleanliness, and maintenance), dietary factors, and recreational activities. Information about a facility is obtained by nurse/social worker teams, who take 1 to 2 days to observe a facility, interview managers and staff, and examine a sample of residents. Some QAI dimensions are predictably related to independently assessed deficiencies in nursing homes, which suggests that the QAI may have promise as a regulatory tool.

The Multiphasic Environmental Assessment Procedure

In subsequent chapters we describe the development of the Multiphasic Environmental Assessment Procedure (MEAP), which incorporates many features of earlier efforts to assess group residential settings for older people. This measure can be applied to a range of settings, from skilled nursing facilities to semi-independent apartments; however, it is also designed to measure fine enough detail to discriminate among facilities of a given type. The MEAP assesses four aspects of the facility environment: resident and staff characteristics, physical features, policies and services, and social climate.

The MEAP comprises five instruments:

1. The Resident and Staff Information Form (RESIF) uses information about residents and staff to measure six aspects of the suprapersonal environment.
2. The Physical and Architectural Features Checklist (PAF) relies on direct observation to obtain information about the physical setting.
3. The Policy and Program Information Form (POLIF) focuses on the policies and services in a setting as reported by the administrator or other responsible staff members.
4. The Sheltered Care Environment Scale (SCES) assesses residents' and staff members' perceptions of seven characteristics of a facility's social climate.
5. The Rating Scale (RS) obtains outside observers' assessments of the physical characteristics of a facility and of the residents' and staff members' functioning.

THE BOOK IN BRIEF

The book is divided into four parts. Part I presents the conceptual framework we used in evaluating group living facilities (chapter 1) and describes our research sample. Our work focuses on panels I and III of Figure 1.1 (the objective environmental system and the social climate); in addition, we operationalized variables from panels II and V (the personal system and resident outcomes), but we did not assess residents' coping responses (panel IV). In chapter 2, we describe a sample of more than 300 facilities drawn from throughout the United States. We provide background information about these facilities and their residents and staff, and we show how the characteristics of resident groups vary among facilities and reflect selection and allocation processes that help to determine where an older person is likely to live.

Part II introduces the MEAP. In chapter 3, we focus on the physical and architectural resources of residential programs. Chapter 4 covers variations in the policies and services provided in residential facilities, focuses on the extent to which behavioral requirements are imposed on residents, and examines the balance between individual freedom and institutional order and stability. Chapter 5 covers the social climates of residential programs, including the quality of relationships, the personal growth orientation, and maintenance and change of the social system. In chapter 6, we compare proprietary, nonprofit, and publicly owned facilities and examine the relationship between program characteristics and program size.

Part III focuses on how to use information about the residential environment in program evaluation. Chapter 7 shows how the social climate emerges from the overall residential context, the resident and staff characteristics, the physical features, and the policies and services provided. In chapters 8, 9, and 10, we use a model of person–environment congruence to help explain the differential impact of program factors on residents who vary in functional and mental status. Chapter 8 involves facility-level analyses that focus on how policies related to choice and control and social climate factors influence residents' well-being and use of facility services. Chapter 9 involves individual-level analyses that consider how the personal and facility characteristics are related to individual residents' activity involvement. In chapter 10, we describe the relocation of a long-term care facility in which we documented changes in the program environment and in residents' and staff members' use of space and patterns of behavior.

Part IV considers applications of the work for designing and improving residential facilities. Chapter 11 highlights variations in residents' and staff members' preferences for specific physical features and policies in residential care facilities and provides information about the preferences of older people living in their own homes or apartments and those of gerontologists with special expertise in housing issues. Chapter 12 summarizes the findings, draws implications for program evaluation and design, and shows how our procedures can help to monitor program changes and promote program improvement. We close by setting forth important issues for future gerontological research and practice.

2

Characteristics of Residential Programs

In undertaking this work, one of our basic aims was to measure the quality of residential programs for older adults. To pursue this aim, we developed the Multiphasic Environmental Assessment Procedure (MEAP), which assesses residential programs in terms of four sets of dimensions: resident and staff characteristics, physical features, policies and services, and social climate. The MEAP is an integrated, conceptually based environmental assessment procedure. After describing the samples of facilities we studied, we focus on the development of the MEAP, on the characteristics of the residents and staff in the facilities, and on selection and allocation issues.

THE SAMPLES OF FACILITIES

We assessed two samples of residential facilities for older adults: a national sample of community facilities and a sample of facilities serving veterans.

Community Facilities

The national sample of community facilities was obtained in two phases. In the first phase, we assessed 93 facilities drawn from several counties in northern California; the facilities varied in locale (urban, suburban, rural) and ownership. Each facility had at least 10 residents, a majority of whom were 60 years of age or older, and offered a meal plan for residents; most of the facilities also offered other health and daily living services. Based on the level of care and services provided, we distinguished three types of facilities: nursing homes, residential care facilities, and congregate apartments.

In the second phase, the sample was expanded to include an additional 169 community facilities from 20 states, representing the major geographic regions of the United States. In general, we followed the same three criteria as in the California sample: Each facility had to have a minimum of 10 residents, a majority of these residents had to be 60 years of age or older, and the facility had to offer a meal plan. The MEAP appeared to be applicable to a broader range of facilities, however, so we included some residential care settings with fewer than 10 residents and some congregate apartments that did not offer a meal plan. (For details of the sampling procedures, see Moos & Lemke, 1992a.)

Because of similarities in characteristics, the facilities from California and from other states were combined into a single normative group of 262 community facilities, which includes 135 nursing homes, 60 residential care facilities, and 67 congregate apartments. Table 2.1 summarizes information on the size and staffing of these facilities and on the demographic characteristics of their residents.

Table 2.1 Characteristics of facilities, staff, and residents in three types of community facilities

	Nursing homes (*N* = 135 facilities)		Residential care *N* = 60 facilities)		Congregate apartments (*N* = 67 facilities)	
	Mean	*SD*	Mean	*SD*	Mean	*SD*
FACILITY CHARACTERISTICS						
Size (number of residents)	104	60	80	84	194	88
STAFF CHARACTERISTICS						
Number of staff per 100 residents	72	17	41	19	7	7
Women (%)	88	8	76	23	48	22
Over age 50 (%)	18	14	29	28	33	24
Employed more than 1 year (%)	56	23	68	24	63	29
RESIDENT CHARACTERISTICS						
Average age (years)	79	5	79	6	76	4
Women (%)	69	17	68	25	76	13
Marital status (%)						
Never married	17	12	18	17	9	7
Married	14	10	7	8	15	9
Divorced	8	8	11	15	11	11
Widowed	61	19	65	22	66	15
Nonwhite (%)	9	14	7	17	20	31
In residence more than 1 year (%)	61	21	71	20	80	24
Receiving public assistance (%)	58	30	25	32	28	28

Nursing Homes

The 135 nursing homes are settings in which professional nursing care is available to all residents. Most of these nursing homes are in one- or two-story buildings. Almost half are in residential neighborhoods, the remainder in mixed use or mainly business neighborhoods. Reflecting the boom in nursing home construction during the 1960s, over half the nursing homes were built during that decade.

On average, the nursing homes have just over 100 residents. They have a high staffing level (72 full-time staff for every 100 residents). In respect to their size and staffing level, these nursing homes are similar to the certified facilities (intermediate care or skilled nursing facilities) included in the 1985 National Nursing Home Survey, which averaged 99 beds and a 92% occupancy rate and had 74 staff per 100 beds (Strahan, 1987).

Among residents, women outnumber men by about two to one; a majority of residents are widowed, and just over half receive public assistance. Only about 9% are from ethnic minority groups. Resident turnover is relatively high; only 61% of the residents have been in the nursing home for more than a year. Most of the staff are women, and their turnover rate is even higher. Only about half the staff in these community nursing homes have worked in the setting for more than a year.

Residential Care Facilities

The 60 residential care facilities provide a moderate level of supervision and assistance with daily living tasks but only limited health services. These facilities are located in older buildings; 52% were constructed before 1960. Many are two- or three-story buildings; just over half are located in residential neighborhoods and the remainder in mixed residential and business areas.

On average, residential care facilities have 80 residents, 24 fewer than the average nursing home, and these facilities are also more varied in size. The staffing level is 41 staff members for every 100 residents. As with nursing homes, the majority of staff members are women, but the turnover rate is lower; about two-thirds of the staff have worked in the facility for more than a year.

In many respects, the residents are similar to those in nursing homes. A majority are women, more than half are widowed, and only 7% are from ethnic minority groups. In contrast with nursing homes, the resident population is somewhat more stable; about 70% of the residents have lived in the facility for more than a year. Reflecting the lower cost of these settings, fewer residents than in nursing homes receive public assistance.

Congregate Apartments

The 67 apartment facilities offer minimal services—typically at least a meal program but not routine care to the majority of residents. Of the three facility types, they are most likely to be located in neighborhoods with some businesses (63%). As a result of federal housing programs, the construction of congregate apartments for elderly people was concentrated in the early 1970s; half were built in the years from 1970 to 1976. They are generally high-rise structures housing an average of almost 200 residents.

The staffing level is low, reflecting the minimal level of direct services offered. Unlike the other types of facilities for older people, the staff in apartments is evenly divided between men and women. Staffing is somewhat less stable than in residential care facilities; 63% have worked in the facility for more than a year.

On average, residents of congregate apartments are slightly younger than residents of the other types of facilities in this sample. The apartments have a higher proportion of women and widowed persons and are also more likely to attract minority elderly people than are the other types of settings. A total of 80% of the apartment residents have lived in the facility more than a year.

Overall Characteristics of Community Facilities

This normative sample, which was drawn to reflect the diversity of group residential settings available to older adults, is not a random or strictly representative sample. However, through discussions with individuals involved in patient placement and facility regulation, we made an effort to include facilities that were considered to provide good and poor quality care. The variation among facilities in physical features, policies and services, and social climate indicates that this effort was successful (see chapters 3, 4, and 5).

Compared with the population of persons over age 60, residents in these congregate living situations are older, more likely to be women, and more likely to be widowed. Stability of the resident population is inversely related to level of care, with apartments having the highest proportion of long-term residents and nursing homes the lowest. The makeup of the staff is also related to level of care. Facilities providing more services tend to be staffed by women, whereas those with the fewest services tend to draw equal numbers of men and women staff. The nursing homes experience the highest staff turnover. Although residential care facilities are like nursing homes in that they are staffed mainly by women, they tend to have the most stable staffing pattern. Apartments are intermediate in the number of longer-term staff.

Veterans Facilities

We also applied the MEAP to a sample of 81 facilities that serve veterans, a relatively unstudied group of facilities. We obtained this sample in order to compare public with proprietary and nonprofit facilities and to increase the diversity of the overall sample in such aggregate resident characteristics as age and sex. Like the community sample, these facilities are drawn from various regions of the United States, covering a total of 36 states. Table 2.2 summarizes characteristics of the two subsamples of these facilities and their residents.

Veterans Nursing Care Units

The 57 facilities in this subgroup include 36 nursing home units and 21 long-term care units. A total of 54 of these units are based in Department of Veterans Affairs (VA) Medical Centers and 3 units are in state veterans homes. Just over half the facilities are in residential areas; the remainder have some businesses close by. The facilities are located in relatively old buildings, most of which are low-rise. On average, they house just over 80 residents, though the range of size is quite broad.

These units provide a high level of services, including not only 24-hour nursing care but other medical services as well. Consequently, the staffing level is high (84 full-time staff for every 100 residents). A majority of the staff members are women. The staffing is quite stable; almost three-fourths of the staff have worked in their unit for more than a year.

Nearly all residents in these units are men, with an average age of 71 years. More than a third of the residents are currently married; the remainder is divided among the categories of widowed, divorced, and never married. About 10% are members of ethnic minority groups. The population of these units is quite stable; almost two-thirds of the residents have been there for more than a year.

Table 2.2 Characteristics of facilities, staff, and residents in two types of veterans facilities

	Nursing care units (N = 57 facilities)		Domiciliaries (N = 24 facilities)	
	Mean	*SD*	Mean	*SD*
FACILITY CHARACTERISTICS				
Size (number of residents)	82	83	128	92
STAFF CHARACTERISTICS				
Number of staff per 100 residents	84	25	32	15
Women (%)	70	15	55	15
Over age 50 (%)	26	12	36	25
Employed more than 1 year (%)	74	20	85	17
RESIDENT CHARACTERISTICS				
Average age (years)	71	4	68	6
Women (%)	4	7	12	27
Marital status (%)				
Never married	28	12	28	14
Married	38	12	13	15
Divorced	18	10	35	19
Widowed	17	10	23	19
Nonwhite (%)	11	10	6	9
In residence more than 1 year (%)	65	21	74	14

Domiciliaries

The 24 domiciliaries include 11 VA-run domiciliaries and 13 domiciliary sections in state veterans homes, about equally divided in location between residential and business neighborhoods. Like community residential care facilities, domiciliaries are designed to provide supervision and assistance with daily living tasks. The buildings housing these units are older on the average than the other types of veterans facilities. They are also larger, with an average of 128 residents, and quite varied in size.

The domiciliaries provide fewer services than do the other veterans facilities, and this is reflected in the lower staffing level (32 staff for every 100 residents). The staff draws about an equal number of men and women and has little turnover.

Domiciliary residents are somewhat younger than residents in the other veterans settings, but like them, nearly all are men. Compared with nursing care units, domiciliaries have fewer residents who are married. The proportion of widowed and never-married persons is similar in the two types of veterans settings, but domiciliaries have many more divorced individuals. Six percent of the domiciliary residents are minorities. Like veterans nursing care units, the resident population is relatively stable.

Comparison of the two types of veterans settings with the population of men over age 60 shows that the age distribution is quite similar. But they are unlike the general population of older men in marital status; more residents of the veterans facilities are in the never-married and divorced categories and fewer are in the category of presently married, which in the general population encompasses more than 80% of men over age 60.

THE MULTIPHASIC ENVIRONMENTAL ASSESSMENT PROCEDURE

Two major phases of research were involved in developing the Multiphasic Environmental Assessment Procedure (MEAP). A preliminary version was constructed using data from the 93 community facilities in northern California. This version was revised on the basis of data from 151 additional community facilities drawn from various parts of the United States. Finally, to increase the diversity of the sample and to permit comparisons of different ownership types, the MEAP was applied to 18 additional nonprofit facilities and to 81 facilities for veterans.

The MEAP consists of five instruments, which can be used either separately or in conjunction with one another. The content of each of the first four instruments reflects one of the four environmental domains identified in Chapter 1.

1. The Resident and Staff Information Form (RESIF) measures the suprapersonal environment. It is composed of six dimensions describing the residents' social backgrounds, functional abilities, and participation in activities as well as the characteristics of staff and volunteers.

2. The Physical and Architectural Features Checklist (PAF) assesses eight dimensions and covers questions about the facility's location, external and internal physical features, and space allowances.
3. The Policy and Program Information Form (POLIF) assesses nine dimensions, including questions about the types of rooms or apartments, facility rules and policies, and the services provided for residents.
4. The Sheltered Care Environment Scale (SCES) assesses residents' and staff members' perceptions of seven aspects of a facility's social climate. The items cover the quality of interpersonal relationships, the opportunities for personal growth, and the mechanisms for system maintenance and change.
5. The Rating Scale consists of four subscales that measure observers' evaluative judgments of the physical environment and of resident and staff functioning.

These instruments allow an evaluator to obtain information about a facility from a variety of sources, including direct observations of physical features and of the residents and staff, interviews with administrators and other staff, review of facility records, and the residents' and staff members' reports of the facility's social environment. Taken together, these diverse sources of data provide an integrated picture of the quality of a facility.

Preliminary Version of the MEAP

We took a number of steps to develop the preliminary items and dimensions of the MEAP. We made observations in group living settings; interviewed residents, staff, and administrators; had discussions with state and local government inspectors; and conducted a thorough search of the literature. To ensure that items were clear and relevant to the range of residential facilities for older adults, we pretested specific items and various response formats with residents and staff. In principle, each item had to apply to nursing homes, residential care facilities, and congregate apartments. Our goal was to develop a common scoring key to permit comparisons between the three types of facilities.

Based on data from the California facilities, we used four criteria to develop a scoring key: (1) Items had to be conceptually related to a specific dimension to be placed on it; (2) each item had to have a varied distribution in the sample; in addition, where appropriate, we selected items that discriminated between different types of facilities; (3) we tried to develop subscales with a moderate to high level of internal consistency; that is, the items on each dimension had to be empirically related to each other; and (4) to increase conceptual clarity and minimize subscale overlap, we placed each item on only one dimension. We collapsed subscales that were conceptually similar and highly interrelated to reduce redundancy.

Revision of the MEAP

In the next phase of the project, we applied the preliminary version of the MEAP to a group of 151 community facilities. In light of experience with additional data collection and provision of results to participating facilities, we reviewed items for their face validity on a particular dimension. Using the national sample data, we again examined the empirical interrelatedness of items on a dimension. Finally, the subscales were intercorrelated to examine whether any of the dimensions could be combined in order to increase subscale independence.

These procedures resulted in the four main parts of the MEAP and the Rating Scale. In this chapter, we describe the aggregate characteristics of residents and staff in the samples of facilities, based on information obtained with the RESIF. The following chapters focus on the physical features (chapter 3), policies and services (chapter 4), and social climate (chapter 5) of the community facilities. Each chapter covers the conceptual rationale for the relevant dimensions and presents normative and descriptive information about the three groups of community facilities. The *MEAP User's Guide* provides more information about the data collection and MEAP scale-construction procedures (Moos & Lemke, 1992a).

THE SUPRAPERSONAL ENVIRONMENT OF RESIDENTIAL SETTINGS

When individuals come together in a social group, be it a college dormitory, a work group, or a congregate living unit for older people, they bring with them values, norms, and abilities. Because of selective mechanisms, groups, which draw their members in a nonrandom manner from the general population, produce distinctive blends of these individual characteristics. The aggregate of residents' and staff members' attributes (the suprapersonal environment) in part defines the subculture that develops in a group; in turn, this subculture influences the behavior of individual members. Hochschild's account (1973) of the community that evolved among the rural-born, white, working-class, widowed women of Merrill Court aptly illustrates this process in a congregate apartment (see also Ross, 1977).

Although older people are sometimes thought of as a stereotypic and homogeneous group, individual differences are maintained or even increase with age (Nelson & Dannefer, 1992; Rowe & Kahn, 1987). But self-selection and social allocation tend to make settings more homogeneous than the overall population. People who share attributes with a group are more likely to select that group. In addition, factors such as referral source, geographic location, cost of care, and reimbursement procedures operate to increase the likelihood of an individual's placement in a particular residential facility (Kane & Kane, 1987).

As a result of the diversity among older people and these selective mechanisms, settings tend to differ in the average characteristics of their residents, that is, in the suprapersonal environment. The features of the suprapersonal environment may be especially salient for older residents of group residential facilities. In general, older adults are more likely to interact with their immediate neighbors than with those who live at a greater distance, even closest friends (Adams, 1985–86). Moreover, because many elderly people in residential settings are isolated from community roles and activities, they are more dependent on their living setting and the other people in it. The reduced functional ability of some older people may also make them susceptible to the influence of the immediate suprapersonal environment.

The mix of residents can affect the quality of life in a setting by influencing staff perceptions of residents' behavior. Morgan (1985) noted that nurses' perceptions of residents' mental confusion were influenced by residents' social characteristics and by characteristics of the facility as well as by residents' medical conditions. In facilities with younger residents, nurses were more likely to rate residents as mentally confused, even after controlling for their objective levels of confusion. Such stereotypes may affect facility policies and services.

Similar processes of selection apply to the staff in a facility. Some facilities may be more likely to attract older staff members, men, or staff with more training. Such differences in the aggregate characteristics of staff may influence the organization of the facility, its social norms, and the program emphasis.

These considerations underline the importance of including aggregate resident characteristics in any examination of group living facilities. Using this information, a number of issues can be addressed. How much do group settings for older people vary in aggregate characteristics of their residents and staff? Are residents sorted on the basis of their personal characteristics into settings that offer different environmental resources? Do residents develop distinctive activity patterns in different settings?

The Resident and Staff Information Form (RESIF)

We assessed the aggregate characteristics of residents in each facility with the RESIF, which includes 69 scored items. To complete the RESIF, we obtained information on background characteristics from records or from residents' direct reports. Information on residents' functional abilities and activity patterns generally came from a staff member directly involved with this area of residents' lives. For example, in nursing homes, the nursing staff provided information on functioning, and the activity director provided data on participation in organized activities. In some apartments, residents reported on their own functioning and activity level.

Dimensions of Suprapersonal Resources

In order to make information from the RESIF useful to practitioners and researchers, we organized the items into the six dimensions shown in Table 2.3. Two dimensions are based on the sociodemographic characteristics of residents. Resident Social Resources assesses the social competence of residents as reflected by their marital status, education, former occupation, and reliance on public assistance. Resident Heterogeneity measures the extent to which residents are a diverse group of individuals with respect to sociodemographic factors, such as gender, age, ethnicity, education, and religion.

Table 2.3 Resident and Staff Information Form (RESIF) subscale descriptions and item examples

RESIDENTS' SOCIODEMOGRAPHIC CHARACTERISTICS	
1. Resident Social Resources (4 items)	Current status of residents with respect to demographic variables that facilitate social competence (What proportion of residents are married? What proportion have a high school education?)
2. Resident Heterogeneity (10 items)	Extent to which residents are a diverse group of individuals (What proportion of the residents are male? What proportion come from various religious backgrounds?)
RESIDENTS' ABILITY AND ACTIVITY LEVELS	
3. Functional Abilities (15 items)	Residents' independence in performing daily functions and the number of handicaps in functioning (What proportion of residents can take care of their own appearance? What proportion do not know what day or year this is?)
4. Activity Level (13 items)	Extent to which residents are involved in activities that they initiate themselves (What proportion of residents read a newspaper or book in the past week? What proportion of residents took a walk during the past week?)
5. Activities in the Community (14 items)	Residents' participation in activities that take them outside the facility (What proportion of residents go out to attend a concert or play? What proportion of residents engage in volunteer or paid work?)
STAFF CHARACTERISTICS	
6. Staff Resources (13 items)	Resources available from staff in terms of experience, training, and variety of backgrounds (Is there a doctor and an occupational or physical therapist? Is there an in-service training program?)

The next three dimensions measure aspects of residents' current functioning. Resident Functional Abilities assesses residents' ability to perform without assistance such daily functions as grooming, eating, dressing, walking, and bathing and the residents' freedom from handicapping conditions. Activity Level focuses on the extent to which residents are involved in activities that they can initiate themselves, such as watching television, sewing or knitting, and visiting with other residents. Activities in the Community is a measure of resident participation in activities outside the facility, such as going out to visit friends or relatives, going shopping, or attending religious services.

Finally, Staff Resources evaluates the staff in terms of their experience, training, and heterogeneity of background. (For the specific items on each dimension and more information about the normative and psychometric characteristics of the RESIF, see Moos & Lemke, 1992e.)

Variations among Types of Facilities

Figures 2.1 and 2.2 show the average scores on the six RESIF dimensions for the community and veterans facilities. The average scores for two dimensions (Resident Heterogeneity and Staff Resources) reflect the percentage of applicable items in the scored direction. Scores for the four other dimensions reflect the average proportion of residents who share a given characteristic or who engage in an activity at specified levels of frequency. For example, Activities in the Community is composed of 14 items; facility scores can vary from 0% (none of the residents leave the facility for any activities) to 100% (all the residents go out for each of the activities). The mean score of 15% obtained for the 135 nursing homes on Activities in the Community (see Figure 2.1) indicates that, on average, 15% of the residents are involved in the outside activities we surveyed.

Community Sample

The three facility types differ somewhat in the demographic characteristics of their residents (Figure 2.1). On average, residents in nursing homes have fewer social resources compared with residents in residential care and apartment facilities. Although they do not vary much with facility type, Social Resources and Resident Heterogeneity vary substantially among facilities within each of the three subgroups of facilities.

The facility types show more striking differences for the next three dimensions. Functional Abilities, Activity Level, and Activities in the Community are highest in apartments, intermediate in residential care facilities, and lowest in nursing homes. Apartment residents were able to function independently in more than 90% of the abilities assessed, which indicates that they are similar in level of impairment to older people who live in the community (Wiener et al., 1990). In contrast, nursing home residents were independent in only 37% of these areas. Our sample of nursing homes is quite similar in this respect to the nursing homes in-

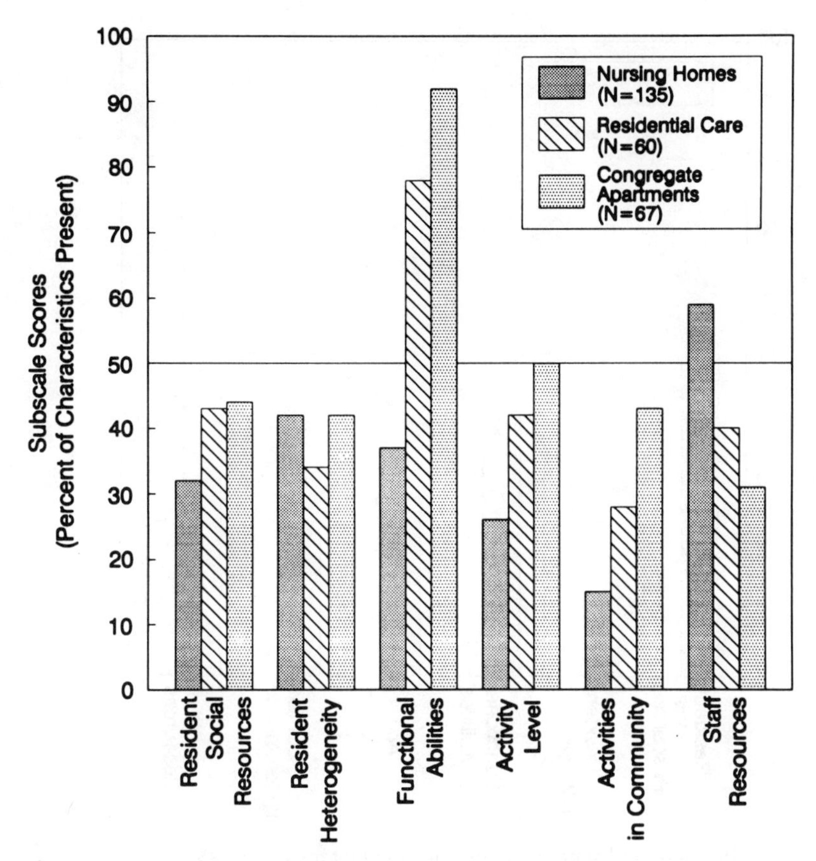

Figure 2.1 Resident and staff characteristics in community facilities.

cluded in the 1985 National Nursing Home Survey (NNHS). In our sample, 87% of nursing home residents require assistance with bathing (vs. 90% in the NNHS), 70% require help dressing (vs. 76%), 41% need assistance with eating (vs. 38%), and 61% are sometimes incontinent (vs. 51%; Wiener et al., 1990).

Because almost all residents in congregate apartments can function independently in tasks of daily living, the apartments show little variability on the Functional Abilities dimension. However, resident groups in the apartments vary on Activity Level and Activities in the Community. When used together, these three subscales can tap different aspects of functioning and activity involvement for residents in all three types of facilities.

Paralleling differences in resident functioning, staff–resident ratio, and level of services, Staff Resources are highest in nursing homes, intermediate in residential care facilities, and lowest in congregate apartments. In addition to the differences between facility types, Staff Resources vary substantially within each of the three types of facilities.

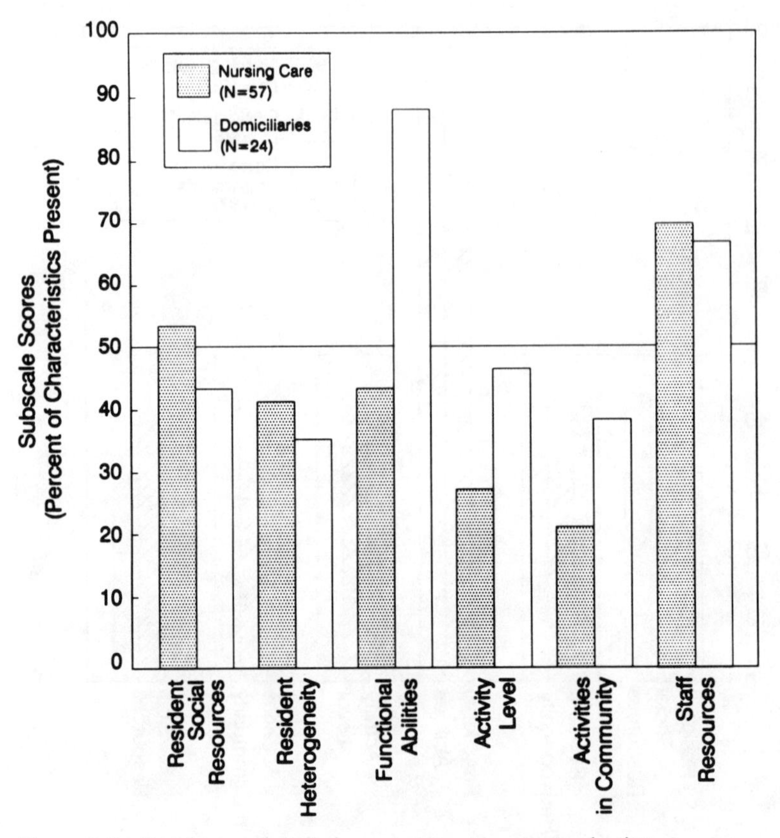

Figure 2.2 Resident and staff characteristics in veterans facilities.

Veterans Sample

Figure 2.2 shows the means of the six RESIF dimensions for the 57 veterans nursing care units and the 24 domiciliaries. Nursing care unit residents have more social resources than do domiciliary residents, although the two groups are comparable in their heterogeneity. Compared with nursing care unit residents, domiciliary residents are much less impaired, have a higher activity level, and engage in more activities in the community. The different types of veterans facilities provide comparable Staff Resources. Both the nursing care facilities and domiciliaries show substantial variability on most of the RESIF dimensions.

SELECTION AND ALLOCATION

We have found wide variations in the characteristics of resident populations in different facilities. These differences reflect mechanisms of selection and social allocation that determine where a person is likely to live.

The inhabitants of residential settings do not constitute a simple cross section of the older population, and they are not randomly distributed into available programs.

A variety of personal and social forces helps determine who seeks congregate residential options and the level of care and specific facilities where they live. Andersen and Newman (1973) have conceptualized these forces as need, predisposing, and enabling factors. Need factors are functional and cognitive deficits that reflect a potential need for nursing home or supportive residential services. Predisposing factors, such as gender and age, affect individuals' propensity to use nursing home or other residential services independent of their need for care. Finally, enabling factors help individuals control whether or not they will use residential care services and, if so, the level of services. These factors include socioeconomic status, married status, and informal caregiver support (Andersen & Newman, 1973; Greene & Ondrich, 1990; Mutran & Ferraro, 1988).

These factors continue to operate after an individual becomes a resident of a congregate facility. For example, individuals who live alone or do not have a caregiver are more likely to enter and remain in a residential facility. Although there are some exceptions, these factors tend to increase the homogeneity of resident populations and the match between resident and facility characteristics. We examined these factors in our samples to learn more about selection into congregate facilities and into different levels of care.

Selection into Residential Settings and into Different Levels of Care

Need Factors

A number of studies have identified factors that predict entry into a congregate residence and into different levels of care. In general, need for care is most important. As a group, applicants and residents of nursing and residential care homes are more functionally and mentally impaired than community samples of people over age 60, and nursing home residents are more impaired than residents of other types of facilities (Branch & Jette, 1982; Branch & Ku, 1989; Braun, Rose, & Finch, 1991; Cohen, Tell, & Wallack, 1986; Mor et al., 1986).

In our sample, functional ability is an important selection criterion into long-term care and into different levels of care. As noted earlier, residents in nursing homes and residential care facilities are more impaired on average than are older community residents. In addition, residents in different levels of care in both the community and veterans facilities show marked differences in their current functioning and activity levels.

Like older people in the community (Elston, Koch, & Weissert, 1991; Wiener et al., 1990), fewer than 10% of apartment residents need assistance with everyday tasks such as dressing, bathing, eating, and shopping. They are relatively active and participate frequently in activities outside the facility that bring them into contact with the surrounding community (Figure 2.1). Older adults living in residential care facilities have somewhat lower functioning and activity levels. Nursing home residents are most impaired and least likely to engage in leisure activity. Comparable differences in functional ability and activity levels distinguish the resident populations in veterans nursing care units and in domiciliaries (Figure 2.2).

Predisposing and Enabling Factors

Functional status is an imperfect predictor of entry into care and of level of care, however. The functioning of community residents overlaps that of older people in congregate facilities; some community residents are more impaired than the average nursing home resident, and some nursing home residents are as capable as the average community resident. The overlap is even greater between community and residential care facility residents. This overlap indicates that factors other than functional need contribute to initial entry and later retention decisions. Such predisposing and enabling factors include older age, being widowed or never married, and living alone or not having a caregiver (Branch & Jette, 1982; Cohen et al., 1986; Dolinsky & Rosenwaike, 1988; Greene & Ondrich, 1990; Kemper & Murtaugh, 1991; Wingard, Jones, & Kaplan, 1987).

The overlap in functional status between different levels of care can be traced in part to policies that encourage residential stability. Older people do not change residence each time their functioning improves or deteriorates. A resident of a nursing home or residential care facility may improve in functional status to a level that would not justify initial admission and yet be allowed to continue in residence. Similarly, a lower level of functioning may be permitted among current residents of apartment or residential care facilities than would be permitted for a new admission.

Our findings show that residents in the three sets of community facilities are drawn from roughly comparable social backgrounds, although nursing home residents are less well educated and of lower occupational status than residents in the two other groups of facilities. Higher social status may provide access to resources that help prevent or delay the move to a higher level of care.

We found that residential care facilities have fewer married residents than do nursing homes and apartments, and apartments are least likely to draw residents who have never been married. This finding is consistent with a pattern identified by Kart and Palmer (1987), who noted that long-term care residents who are married and have some family support are less likely to receive personal care in a residential care facility and more likely to receive intensive nursing care that they cannot get in their own home.

Age, per se, is also viewed as a predisposing factor in congregate care, as shown by the development of residential programs for the independent elderly population. Because of frailty and vulnerability, older people may seek relief from the responsibility of managing an independent household and may benefit from the security of a setting where they receive some monitoring and where they can continue to live even if their functioning declines. Within congregate settings, age appears to help determine the level of care in which the older person is placed, as reflected in the greater average age of nursing home and residential care facility residents. Similarly, Cohen and his colleagues (1986) noted that age and its interactions were the best predictor of nursing home entry. For example, they found that with increasing age, a person who needs help to get around is more likely to enter a nursing home. Similar factors have been found to predict older adults' visits to physicians and hospitalizations (Mutran & Ferraro, 1988; Strain, 1991) and use of in-home care services (McAuley & Arling, 1984).

Selection into Specific Facilities within Levels of Care

As we have seen, functional impairment is the major basis for selection of an appropriate level of care. This factor is also involved in determining the choice of a particular nursing home or residential care facility, as indicated by the variation in residents' functioning among each of these two types of settings. As noted previously, apartments show little variation in the functional abilities of their residents.

Such predisposing and enabling factors as educational and socioeconomic status also affect the likelihood of placement in a particular facility. These selection factors give facilities a distinctive character and increase the homogeneity of their residents when compared with the overall population of older adults in residential facilities. In addition, these factors work to provide higher-status residents with access to better-staffed facilities.

Need for Services and Staff Resources

To examine the evidence for selection processes that operate within a given level of care, we conducted hierarchical regression analyses to predict residents' access to staff resources from the level of care in the facility, residents' functional ability, and then residents' demographic characteristics. If older people are matched to settings according to need and if resources are rationally allocated within each level of care, we would expect to find a positive relationship between residents' need for services (as indicated by their functional abilities) and the staff resources available to them.

Two measures of staff resources were included in these analyses: the RESIF dimension of Staff Resources and the ratio of staff to residents. These two measures are only moderately related ($r = .19$ for community facilities and $r = .08$ for veterans facilities). Thus, they tap somewhat separate aspects of staffing characteristics.

Facilities with more functionally impaired residents are somewhat better staffed (partial r's $= .27$ and $.32$ for community and veterans facilities, respectively), but Staff Resources are unrelated to functional ability. Thus, although need for services is strongly related to staffing levels and resources across different levels of care, within a given level of care, groups of more-impaired residents are likely to have only slightly higher staffing levels. Moreover, staff serving groups of more-impaired residents are no better trained, more diverse, or more experienced than those working with less-impaired groups.

Predisposing and Enabling Factors and Staff Resources

Within the levels of care in community facilities, predisposing and enabling factors are as closely associated with variations in staffing as are need factors. After controlling for level of care and functional ability, older people with more social resources live in facilities that have more staff and better-trained staff. More Staff Resources and a higher staffing level are found in settings in which the resident population has higher educational and occupational status. The staffing level is higher where fewer residents receive public assistance and where residents are older; Staff Resources are higher where more of the residents are women.

Veterans facilities show no such relationships between predisposing and enabling factors and the allocation of staff resources. After level of care and functional ability are controlled, there are no significant relationships between residents' social status and staffing characteristics in the veterans facilities. The more egalitarian quality of veterans facilities may grow out of their centralized administration and systemwide staffing guidelines.

Resident–Facility Selection

Resident–Facility Matching Processes

There appear to be two processes involved in resident–facility matching. One process is guided by residents' need for services, as indicated by their functional ability; the other reflects predisposing and enabling factors that enable the more privileged residents to live together and obtain access to better staff resources.

Among the variables we examined, functional ability is the strongest determinant of the type of facility individuals enter and the level of care they receive. Within each level of care, however, some community facilities tend to serve a population that consists mainly of older, widowed women of higher educational and socioeconomic status. In contrast, other community facilities serve residents who are younger, have lower educational and occupational status, and are more likely to be on public assistance.

Once an individual enters the appropriate level of care, these predisposing and enabling factors are as important as need factors in providing access to staff resources in community facilities. Groups of residents who are more dependent have a higher staffing level, but better-trained staff are not allocated to these groups.

Resident Homogeneity

Resident selection and allocation processes result in somewhat homogeneous suprapersonal environments in congregate facilities; that is, residents tend to be more similar to one another in social background and current functioning than are older people living in the community.

Such homogeneity can have both positive and negative consequences. Similarity in cultural and religious background may enhance resident interaction and satisfaction (Bergman & Cibulski, 1981). Age-homogeneous settings may foster social activity and higher morale, in part because they reduce intergenerational role conflict (Hinrichsen, 1985; Rosow, 1967).

By decreasing the diversity of residents' daily experiences, homogeneous settings may not provide enough stimulation and may fail to maintain maximum levels of cognitive and behavioral functioning (Cohen et al., 1987). Selection policies that increase the heterogeneity of individuals in a facility can create a more stimulating environment, which may be therapeutic for some people. For example, Kahana and Kahana (1970) found that older psychiatric patients randomly assigned to an age-integrated program improved in cognitive functioning as much as comparable patients in an intensive therapy program, whereas older patients assigned to an age-segregated program showed no improvement. The age-integrated program apparently provided a more cognitively complex and involving milieu.

The multilevel facility or continuum-of-care model allows for a diversity of residents' needs. In 18-month and 9-year follow-ups, Gutman (1978, 1988) found that functionally able older people who moved into a multilevel facility experienced no declines in self-reported health status, activity level, or social interaction compared with tenants in traditional retirement housing. Although some older people are opposed to sharing space or activities with individuals who are more impaired, a policy of "geriatric mainstreaming" (Hiatt, 1982) may provide egalitarian and stimulating residential facilities that can benefit some residents and staff.

More generally, there may be an optimal level of diversity in residents' functional abilities and service needs. Sherwood and her colleagues (1981) noted that a mix of high- and low-need individuals can improve staff morale and foster mutual-assistance networks among residents. They recommend reserving a proportion of new admissions in sheltered housing facilities for persons with a somewhat lower level of need for this type of environment. Willcocks, Peace, & Kellaher (1987) made a similar plea for reducing the emphasis on homogeneous resident populations and on matching residents with facility types.

In subsequent chapters, we examine the relationships between the aggregate characteristics of residents and such facility features as the physical resources, policies and services, and social climate. These data permit us to analyze further the impact of selection and social allocation on the resources available to older people in congregate residential settings.

II

UNDERSTANDING THE PROGRAM ENVIRONMENT

3

Physical and Architectural Features

The physical design of group living facilities affects the behavior and well-being of older adults. For example, Carp (1987b) noted that positive physical qualities of congregate housing can help to improve older adults' activity levels, social contacts, and well-being; residents of physically attractive congregate housing are also more satisfied (Butterfield & Weidemann, 1987). Similar results have been obtained in long-term care settings (Braun, 1991; Linn et al., 1977). Physical features typically act in conjunction with organizational and social factors to influence resident outcomes (for reviews, see Cohen & Weisman, 1991; Moore, Tuttle, & Howell, 1985; Regnier & Pynoos, 1987).

By evaluating the impact of changes in environmental design, researchers have shown how physical features can influence residents and staff. Osterburg (1987) found that residents in a retirement home reacted favorably to a redesigned building whose organization made it easier for them to identify distinct parts within a unified whole. Residents could more easily orient themselves, and many residents who formerly had meals in their rooms ate in the dining area instead. Lawton (1981) studied how remodeling four-person rooms into private bedroom areas with a central open space affected mentally impaired patients. Residents spent more time outside their bedrooms and became more interested in the immediate environment; their range of movement increased, and pathological behavior and fixed staring declined.

These studies highlight some of the ways in which physical features can influence residents' morale and functioning. However, few methods are available to assess such features systematically. Salient physical features include the location of the facility and accessibility to community services, amenities that enhance residents' comfort, features that foster on-site social interaction and recreation, supportive design features for physically disabled residents, aids to spatial and temporal orientation, security and safety features, and space allowances for resident and staff functions (Gelwicks & Dwight, 1982; Green et al., 1975; Koncelik, 1982; Lawton, 1975). In this chapter, we describe the Physical and Architectural Features Checklist (PAF), which measures these characteristics of group living facilities.

ASSESSING PHYSICAL AND ARCHITECTURAL FEATURES

The Physical and Architectural Features Checklist (PAF)

The PAF uses direct observation of 153 individual items to measure relevant design factors. To make this information more useful to practitioners and researchers, we organized it into eight dimensions (Table 3.1).

Community Accessibility measures the proximity of the facility to resources in the surrounding community. The next two dimensions, Physical Amenities and Social–Recreational Aids, focus on the presence of physical features that add convenience and comfort to a facility and that foster social and recreational activities.

Three dimensions assess physical features that aid residents in activities of daily living and in negotiating the facility environment. Prosthetic Aids measures the provision of aids to physical independence and mobility; Orientational Aids measures the presence of features that orient the resident to time, place, and persons; and Safety Features assesses the availability of features to prevent accidents and unwanted intrusions.

The last two dimensions assess the allowance of space for resident and staff functions. Staff Facilities assesses physical features that make the setting more pleasant for staff; we included this dimension because we thought that such features might promote staff morale and performance and contribute to the quality of resident care. Space Availability measures the amount of communal and personal space available to residents. (For information about the normative and psychometric characteristics of the PAF, see Moos & Lemke, 1992b; for a description of the initial version of the PAF, see Moos & Lemke, 1980.)

Physical Features in Different Types of Facilities

Figure 3.1 shows the means of the eight PAF dimensions for the 135 nursing homes, the 60 residential care facilities, and the 67 congregate apartments in the community sample (results for the veterans sample are presented in chapter 6). The means reflect the percentage of applicable items that the facility offers. For example, Physical Amenities is composed of 30 items; facility scores can vary from 0% (the facility offers none of the items) to 100% (it offers all 30 items). A facility that has 15 of the 30

Table 3.1 Physical and Architectural Features Checklist (PAF): Subscale descriptions and item examples

Subscale	Description
DEGREE OF PHYSICAL INTEGRATION	
1. Community Accessibility (16 items)	Extent to which the community and its services are convenient and accessible (Is there a grocery store within easy walking distance? Is there a senior center within easy walking distance?)
PHYSICAL FEATURES TO IMPROVE CONVENIENCE AND COMFORT	
2. Physical Amenities (30 items)	Physical features that add convenience, attractiveness, and comfort (Is the main entrance sheltered from sun and rain? Are the halls decorated?)
3. Social–Recreational Aids (28 items)	Physical features that foster social interaction and recreational activities. (Is the lounge by the entry furnished for resting or casual conversation? Is there a billiard table?)
PHYSICAL FEATURES THAT PROVIDE SUPPORT FOR RESIDENTS	
4. Prosthetic Aids (24 items)	Extent to which the facility provides a barrier-free environment and aids to physical independence and mobility (Can one enter the building without having to use any stairs? Are there handrails in the halls?)
5. Orientational Aids (13 items)	Extent to which the setting provides features that help orient residents (Is each floor or corridor color-coded or numbered? Is a map showing community resources available in a convenient public location?)
6. Safety Features (18 items)	Extent to which communal areas can be monitored; the presence of features for preventing accidents (Is the outside building area well lighted? Are there call buttons in the bathrooms?)
SPACE FOR RESIDENT AND STAFF FUNCTIONS	
7. Staff Facilities (11 items)	Physical facilities that aid the staff and make it pleasant for them to maintain and manage the setting (Are the offices free of distractions from adjacent activities? Is there a staff lounge?)
8. Space Availability (13 items)	Number and size of communal areas in relation to the number of residents; size allowances for personal space (How many special activity areas are there? What size is the smallest per person closet area?)

physical features obtains a score of 50%. The average Physical Amenities score for the 135 nursing homes is 65% (see Figure 3.1).

The means for the three sets of community facilities reflect meaningful variations among types of group living settings; these variations are related to such factors as the impairment of their residents, their staffing level, and the degree of regulation under which they function. For exam-

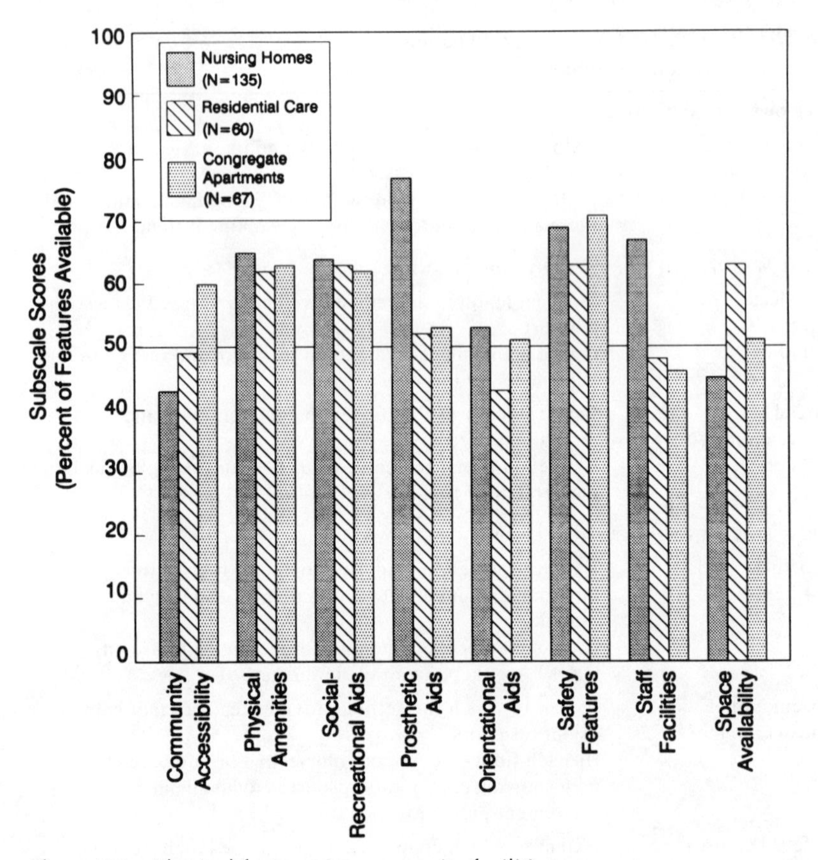

Figure 3.1 Physical features in community facilities.

ple, congregate apartments provide easier access to community resources than do nursing homes or residential care facilities. This difference is due in part to the fact that congregate apartments are more likely to be in high-rise buildings and in more densely developed neighborhoods. The siting of nursing homes and residential care facilities in residential neighborhoods in part reflects planners' and developers' assumptions about the capacities and needs of residents in these types of facilities.

Compared with apartments and residential care facilities, nursing homes offer more prosthetic aids and staff facilities, features appropriate to their populations of impaired residents and high staffing level. In contrast to these differences, the facility types are very similar in the physical amenities and social–recreational aids they offer residents.

Residential care facilities have somewhat fewer orientational aids and safety features than do the other types of facilities. A number of factors may contribute to this finding, including the absence of uniform regulation, the age of their buildings, and their smaller size. On the other hand, residential care facilities provide larger space allowances per resident.

Nursing homes provide residents with substantial communal space but limited personal space, whereas apartments emphasize personal space but provide limited communal space. Residential care facilities appear to be a mixed model with relatively equal emphasis on private and communal space.

Some of the variations between types of community facilities parallel differences in the functioning of their residents, pointing to some matching between residents' needs and the physical resources available to them. Thus, nursing homes have the largest number of impaired residents and the most prosthetic aids. Space for staff functions also is allocated in part on the basis of need. Compared with residential care and apartment facilities, nursing homes have many more staff members; they also have more facilities for staff. On the other hand, access to community resources appears to be allocated, not to compensate for physical disability but on the basis of perceived ability to use these resources.

Specific Aspects of Architectural Design

To provide more information about the design of group living facilities, we can examine the specific physical features available in facilities. We focus primarily on community accessibility and on physical features that contribute to residents' convenience and comfort and those that aid resident mobility and ensure safety.

Community Accessibility

Designers have emphasized that criteria for site selection should include access to frequently used community resources and health and social services. For example, in an evaluation of six public housing sites, Christensen and Cranz (1987) found that about half of the residents visited restaurants, churches, and beauty or barber shops at least monthly and that about 20% used parks, senior centers, and libraries once a month or more. Less attention has been given to the issue of community access for nursing homes and residential care facilities.

In the MEAP, we focused on whether specific community resources are located within easy walking distance (one-fourth of a mile) of the facility. Because they are more likely to be located in residential neighborhoods, nursing homes and residential care facilities have fewer community resources close by than do apartments (Figure 3.1). For example, more than 70% of apartments are within walking distance of a drugstore and a church or synagogue; about half the residential care facilities and nursing homes have these resources nearby. Almost half the apartments have a senior center nearby, but this is true for relatively few nursing homes or residential care facilities.

In spite of the location of apartments in business neighborhoods, some community resources are equally accessible to the different types of facilities. Most have a grocery store nearby, and a majority are near a bank.

Other community resources are generally less accessible. Overall, less than half of the facilities are within walking distance of a park, post office, or movie theater.

In site selection, economic factors and assumptions about residents' needs appear to contribute to more emphasis on community accessibility for apartment facilities. Land costs in neighborhoods with businesses may make high-rise structures the only economically feasible option, and high-rise structures are generally considered less desirable for more impaired residents. In addition, planners may assume that residents of nursing homes will not benefit from community facilities but should have comparable resources (such as movies or religious services) available in the facility.

Physical Amenities

As noted earlier, the three sets of facilities offer similar overall levels of physical amenities. Some features, such as a sheltered entrance, a lawn area, table lamps in lounges, and mirrors in bathrooms are present in more than 80% of facilities. In contrast, less than 40% of facilities offer such amenities as protected outside seating, a chapel, or a gift shop.

The overall similarity in the levels of physical amenities can mask important variations in how different types of facilities allocate their resources. For example, of the three types of facilities, apartments are most likely to provide several resources that support residents' independence, such as access to a kitchen area, access to a laundry area, and individual heating and air-conditioning controls (Table 3.2).

On the other hand, nursing homes, as a consequence of providing less privacy and individual control, offer some amenities that apartments do not. For example, apartments, which have private bathing areas, are less likely than nursing homes or residential care facilities to provide bathtubs, even though most older people, especially women, prefer a bathtub to a shower (Christensen & Cranz, 1987); instead, apartments provide showers, which take less space and are less expensive to install. However, nursing homes and residential care facilities, which usually have shared bathing facilities, often give residents the choice of a bathtub or shower. Nursing homes are most likely to have other communal resources, such as drinking fountains, public phones, and vending machines.

Social–Recreational Aids

Features that support social and recreational activities tend to be communal in nature and are similarly distributed in the three types of facilities. Most facilities have a patio or open courtyard and at least two lounge or activity areas. More than two-thirds of the facilities have a library from which residents can borrow reading material (Table 3.2). Fewer facilities have such features as a snack bar or a music room.

The presence of several of the recreational aids varies in the different types of facilities. For example, nursing homes are least likely to have a

Table 3.2 Percent of facilities responding in the scored direction on selected physical amenities and social–recreational aids

	Percent in scored direction		
Item	Nursing homes ($N = 135$)	Residential care facilities ($N = 60$)	Apartment facilities ($N = 67$)
PHYSICAL AMENITIES			
1. Do residents have access to a kitchen area?	28	55	94
2. Is there a laundry area for residents' use?	16	43	100
3. Is there an air-conditioning system?	64	47	64
4. Are there individual air-conditioning controls?	28	27	54
5. Are there individual heating controls?	39	48	88
6. Do residents have access to both a bathtub and a shower?	97	95	79
7. Are there drinking fountains?	95	56	69
8. Are there public telephones?	90	73	67
9. Are there vending machines that are available to residents?	81	57	62
SOCIAL–RECREATIONAL AIDS			
*1. Is there a patio or open courtyard?	91	90	92
*2. Are there at least two lounge or activity areas?	78	82	91
*3. Is there a library from which books can be borrowed?	68	67	82
4. Is there a quiet lounge with no TV?	57	73	76
5. Is there a billiard table?	23	33	60
6. Is there a radio for general use?	68	82	40
7. Are tables available in an outside area?	75	70	58
8. Is there a telephone connection in residents' rooms?	43	47	100

The three sets of facilities differ significantly ($p < .05$) on all items except those marked with an asterisk (*).

quiet lounge with no television. The absence of this resource appears to reflect ideas about the functioning of residents, specifically, that a television is a necessary resource in all communal lounge areas and that nursing home residents are less likely to use a communal area for activities with which television might interfere. Apartments are most likely to provide a billiard table; the presence of this resource probably reflects assumptions about the level of functioning needed to use this equipment.

Nursing homes and residential care facilities offer some communal resources that apartments do not. For example, they are more likely than apartments to provide a communal television and radio and patio furniture. As a consequence, apartment residents who do not own their own television, radio, or patio furniture are more likely to have to do without these resources. On the other hand, all apartments have a telephone or telephone connection. Fewer than half of the nursing homes and residential care facilities have this option.

Prosthetic Aids

Regulatory requirements, building age, and perceived need of residents are factors related to the presence of prosthetic aids. In general, nursing homes have more prosthetic aids than do either residential care or apartment facilities (Figure 3.1). Most nursing homes have better wheelchair accessibility and more features for the frail elderly, such as reserved parking for handicapped persons, wheelchair-accessible drinking fountains and public phones, and lift bars in bathrooms (Table 3.3). Fewer residential care or apartment facilities have these features, even though residential care facilities are intended to serve those requiring supervision and assistance in daily activities and many congregate apartments were designed to serve both elderly and handicapped populations.

Some prosthetic features are equally common in nursing homes and apartments, a finding that reflects changes in building standards and the relatively recent construction of these facilities. Nearly all nursing homes and apartments have good wheelchair access to the building, including a wide main entrance and the absence of steps. Residential care facilities, in part because of building age and absence of regulation, are somewhat more likely to have entrance steps and a narrow doorway (Table 3.3). In addition, residential care facilities are more likely to require the use of stairs inside the building and to have raised thresholds in the bathrooms; both of these features are obstacles to wheelchair use and to frail elderly people.

As with some amenities, some prosthetic aids are more common in nursing homes because they are offered communally. Prosthetic aids for bathing, such as wheelchair-accessible showers, showers with a flexible shower head, and showers with a seat, are most common in nursing homes and least common in apartments (Table 3.3). Finally, disabilities other than mobility impairment have been given less attention in all facilities (Hiatt, 1987). Although wheelchair-accessible public phones are

Table 3.3 Percent of facilities responding in the scored direction on selected prosthetic aids and safety features

Item	Percent in scored direction: Nursing homes ($N = 135$)	Residential care facilities ($N = 60$)	Apartment facilities ($N = 67$)
PROSTHETIC AIDS			
1. Is there parking reserved for the handicapped?	84	30	39
2. Can one enter the building without having to use any stairs?	95	82	100
3. Is the front door wide enough for a wheelchair?	99	83	99
4. Is the drinking fountain accessible to wheelchair residents?	78	28	29
5. Is at least one public telephone accessible to wheelchair residents?	86	55	45
6. Is a wheelchair-entered shower available?	93	42	7
7. Is there a flexible showerhead?	92	68	37
8. Is there a seat included in the shower?	82	53	27
9. Are there lift bars next to the toilet?	85	60	57
SAFETY FEATURES			
*1. Is there an individual who usually monitors access to the building?	62	63	67
*2. Is there a place for visitors to sign in?	37	33	37
3. Is the outside building area well lighted?	91	78	91
*4. Is the entrance area visible from the lobby or social space?	81	85	85
5. Is the outside seating in front of the building visible from the lobby or social space?	84	68	91
6. Are there call buttons in the bathrooms?	94	41	62

The three sets of facilities differ significantly ($p < .05$) on all items except those marked with an asterisk (*).

fairly common, only about 20% of facilities have phones with a receiver that can regulate the volume of incoming speech for hearing-impaired persons.

Safety Features

Safety features are generally similar across facility types. About two-thirds of the facilities have someone who monitors entry to the building; about one-third have a formal sign-in procedure for visitors (Table 3.3). Although most facilities thus monitor the main entrance, about two-thirds allow access through other entrances, which makes the monitoring procedure less effective. In most cases, the outside building area is well lighted, and the entrance and outside seating can be monitored from the lobby (Table 3.3).

In addition to having features to guard against intruders, these buildings include measures to ensure against accidents to residents, and here again the pattern is somewhat mixed. For example, although facilities with stair areas generally have nonskid surfaces and good lighting, only about half the facilities have nonslip surfaces in bathroom areas subject to wetness. Nearly all nursing homes have call buttons in bathrooms and in individual rooms, but only about half the residential care facilities have call buttons in these locations. Even with a more intact resident population, apartments are somewhat more likely than residential care facilities to have call buttons in bathrooms and individual rooms.

Conclusions

This examination of individual features enriches the picture of how facility types differ in the physical environment they provide residents. Nursing homes provide a communal setting with relatively rich shared resources, particularly those that protect and support residents. The physical environment reflects an orientation toward taking care of residents' needs and a minimal emphasis on the private, independent aspects of residents' lives. On the other hand, congregate apartments support residents' independent functioning. They are well integrated in the community and tend to provide modern surroundings, including some normative prosthetic, orientational, and safety features. But for those apartment residents with limited functional capacity or personal resources, communal features are often lacking.

In contrast, residential care facilities tend to provide less support for the independent functioning of their residents than do apartments and to have many fewer communal resources than nursing homes, particularly in the older and smaller residential care facilities. Both nursing homes and apartments explicitly acknowledge the public facet of the settings with such features as seating in the lobby, a reception desk, and bulletin boards. Residential care facilities are less likely to include these features, thus downplaying further the public, communal side of these settings.

PHYSICAL FEATURES AND THE ADEQUACY OF PROGRAM IMPLEMENTATION

To determine how well a program is implemented, an evaluator must measure the actual program against a standard of what the program should be. Sechrest and his colleagues (1979) identified three sources of information that can be used to define a standard for program implementation: normative data on conditions in other programs, which allow the evaluator to see how one program compares with others; specifications of an ideal program; and conceptual analysis or expert judgment.

The PAF can be used to portray the physical features of a facility and to compare them with normative standards. One useful way to display this information is to profile the results as standard scores. The standard scores (with a mean of 50 and a standard deviation of 10) are based on the appropriate normative sample; for example, a nursing home is compared with other nursing homes. The resulting profiles indicate how well a program is implemented from a normative perspective. They identify areas in which the facility places comparatively more emphasis. To illustrate, we present PAF profiles for three facilities: a nursing home, a residential care home, and a congregate apartment; each is compared with similar facilities in our normative sample. (For additional examples of PAF profiles, see Lemke & Moos, 1985; Moos & Lemke, 1984, 1992b; Moos, Lemke, & Clayton, 1983; Moos, Lemke, & David, 1987; Timko & Moos, 1991a.)

Middleton Nursing Home

Middleton Nursing Home is a large facility located in an urban area in the Midwest. This single-story structure, with a capacity of 175 beds, was built in the late 1960s; it is owned and operated by a proprietary nursing home chain.

Middleton is operating near capacity. Its staff and resident characteristics are fairly typical of nursing homes in the National Nursing Home Survey and similar in most respects to nursing facilities in the MEAP normative sample. Two-thirds of the residents are women. More than half the residents are widowed; 12% are currently married. In these respects, the residents are similar to the general population of nursing homes.

With an average age of 76 years, residents are somewhat younger than residents in our nursing home sample, and fewer (32%) receive Medicaid. In addition, residents are less independent in daily activities than residents in the nursing home sample; on average, only 18% of the residents are independent in a variety of daily living activities. Residents are also less involved in independent social and recreational activities. Middleton Nursing Home has an average level of staffing to serve this impaired population (78 staff for every 100 residents), and they reflect a typical level of experience and diversity.

Figure 3.2 shows the PAF profile for Middleton Nursing Home. Its urban location puts it within walking distance of a large number of community resources. The facility has a slightly above average number of features to make the setting pleasant and to stimulate activity. The physical environment is well adapted for an impaired population, as shown by its relatively large number of prosthetic features and a somewhat above average number of orientational aids. For example, all areas of the building are fully accessible to wheelchairs, handrails and lift bars are appropriately placed, and showers are fitted with flexible heads and seats. In contrast, features to ensure residents' safety are notably absent; in particular, the entry, outside seating, and lobby area are not visible from staff areas or public spaces, which makes monitoring for accidents or intruders difficult. Space is limited in ways that affect residents somewhat more than staff.

Thus, Middleton Nursing Home provides a supportive physical environment with good accessibility to community resources for its relatively young but impaired residents. It may seem somewhat crowded, however,

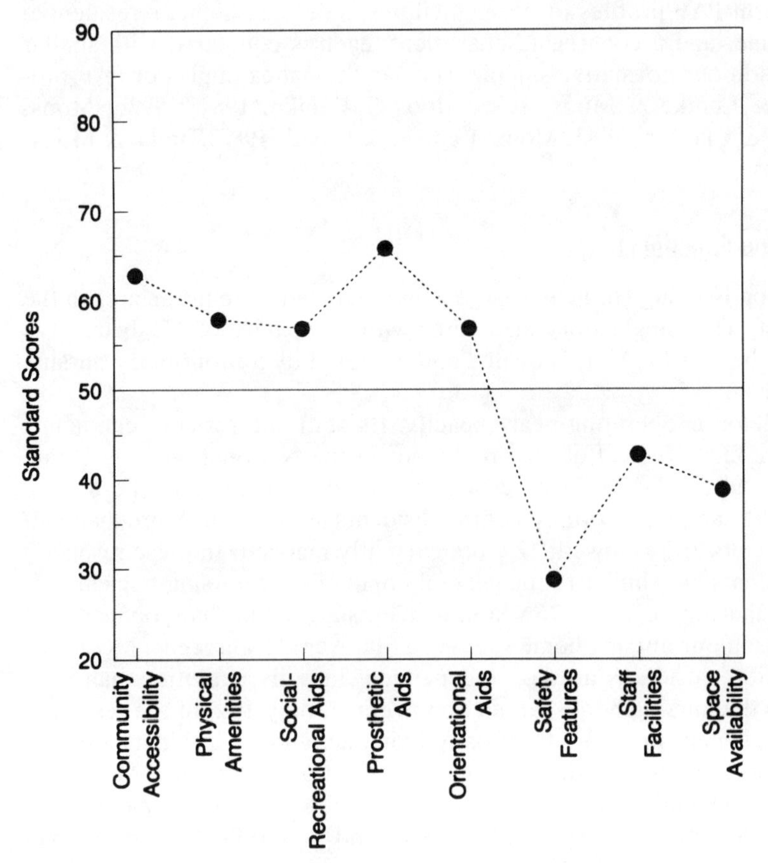

Figure 3.2 Physical features in Middleton Nursing Home.

particularly for residents, and residents may be vulnerable to accidents or intrusion.

Maude's Community Care Home

Maude's is a community care home that houses 17 residents in a two-story building located in a West Coast suburb. It has been run as a family business since the early 1970s, when the home was constructed.

Residents are much less diverse than is generally the case in the residential care facilities in the MEAP sample. Two-thirds are men, most are in the age range from 75 to 84, all but one are widowed, and all are white. They come from a lower socioeconomic stratum. Most completed some high school and held unskilled or blue-collar jobs. Their level of impairment is typical of residential care facilities in the normative sample, but their activity level and involvement in community activities are somewhat lower. Five full-time staff members provide services. Staff Resources are below average; staff members have little formal training and are not very diverse in background.

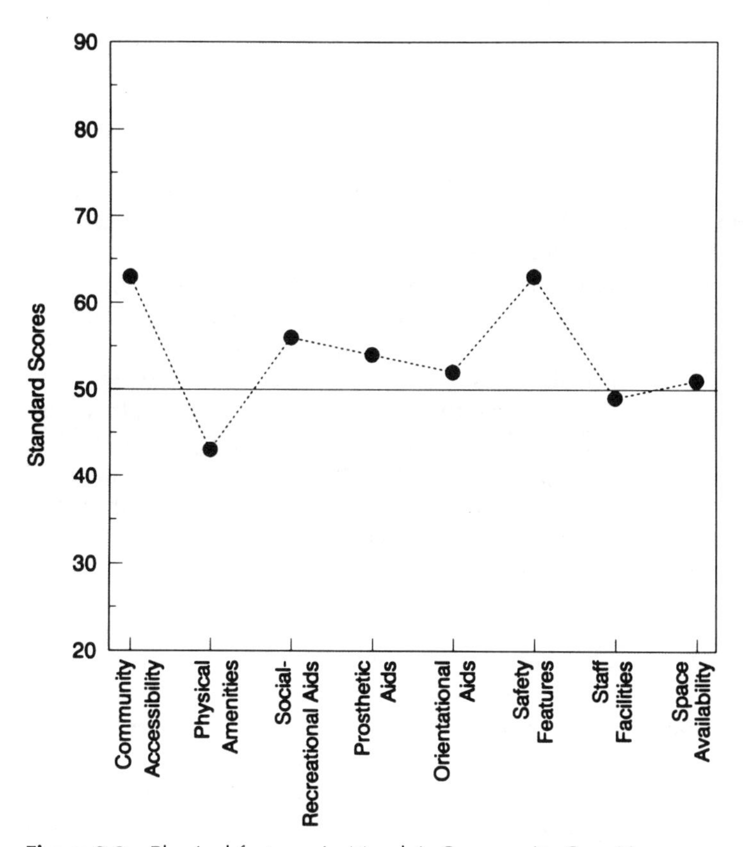

Figure 3.3 Physical features in Maude's Community Care Home.

Figure 3.3 shows the PAF profile for Maude's. This profile compares Maude's with other residential care facilities in the normative sample. Its suburban location puts Maude's within easy walking distance of a large number of community resources. Maude's provides residents somewhat fewer amenities and slightly more social–recreational aids than is typical in residential care facilities. The features designed to support social interaction and recreational activities include outside seating near the entrance, a garden area for residents, small tables and desks in the lounge, a library of reading material, a piano, a pool table, a phonograph, and a radio.

Maude's incorporates a variety of features to create a safe and secure environment. The outside seating and entrance are monitored from staff areas and social spaces, the entrance and stairways are well lighted, and bathrooms have nonslip surfaces and emergency call buttons.

Maude's provides a safe environment with easy access to community resources. In spite of the supports for activity within the facility and in the surrounding community, activity involvement is below average. In other respects, the physical environment is typical of residential care facilities. It has somewhat fewer than average amenities, average prosthetic and orientational aids, and average space allowances for staff and residents.

Amity House

Amity House is an apartment complex constructed in the 1960s in a rural area of the eastern United States. A local religious organization built and manages Amity House under a program of federally assisted housing. This single-story structure houses just over 100 residents. They come from relatively privileged backgrounds; a majority completed college and held professional or executive positions. Less than 10% receive federal rent assistance or a private housing subsidy. Three-fourths of the residents are women, which is typical for congregate apartments in this sample. Residents have a high level of functioning and initiate many activities, both in the facility and in the community. The staffing level of about 27 full-time equivalents is quite high, and the staff is diverse in characteristics. Most staff members work in maintenance or kitchen jobs.

Amity House is very isolated from community services (Figure 3.4). It is in an undeveloped area but residents have access to public transportation. Due in part to its rural location, the apartment complex has fewer than average features to guard against intruders and a lower score on safety features; for instance, the outside seating and main entrance area cannot be monitored from staff offices, several entrances are left unlocked, and no one monitors access to the building.

However, the building does offer residents many amenities and prosthetic aids as well as ample space allowances. The amenities include an air-conditioning system with individual controls in each apartment, a gift shop, a laundry room, and convenient drinking fountains and public

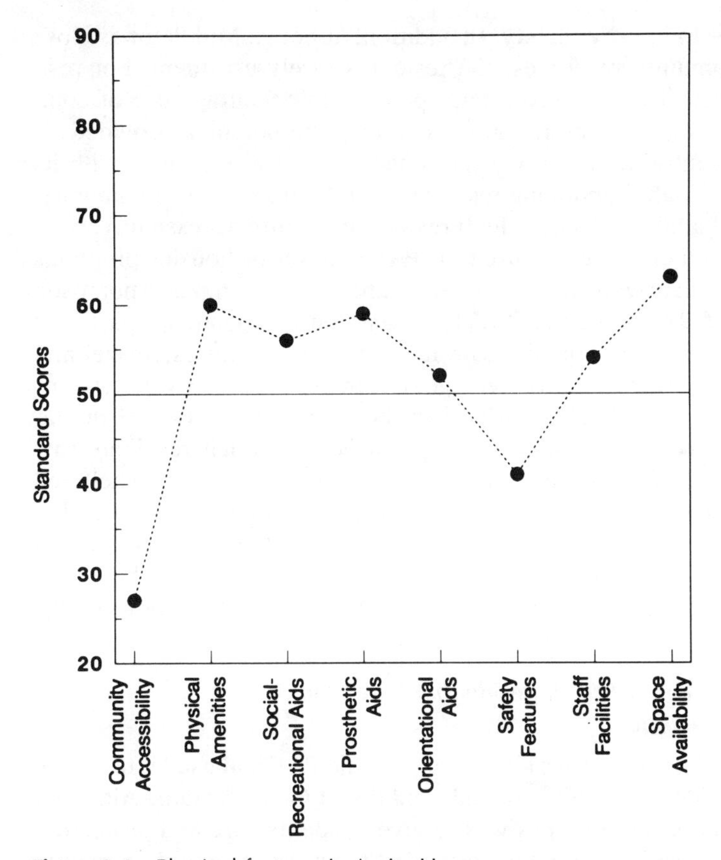

Figure 3.4 Physical features in Amity House.

phones. Among the prosthetic aids less frequently found in apartments, Amity House has reserved parking for handicapped individuals, public phones that are wheelchair accessible, and seats in the showers. In other respects, it ranks about average for apartment facilities.

Program Design and Standards of Implementation

These three examples illustrate some variations in the physical resources of group residential facilities. Middleton provides a supportive environment but lacks safety features and space allowances commonly found in nursing homes. Maude's provides a safe environment with good access to community resources and shows no particularly weak areas when compared with other residential care facilities. Amity House has ample physical resources, but its rural location makes it somewhat isolated from the surrounding community and has perhaps contributed to a more relaxed attitude about some safety features.

According to normative standards, the three programs are relatively well designed and matched with their residents' needs. Given the vulnerability of its residents and its urban location, however, Middleton could make changes to improve safety. In addition, although Middleton is close to many community resources, the residents rarely use them. For residents with their level of impairment, policies to encourage use of community resources may be critical for realizing the potential provided by Middleton's central location. On the other hand, Amity House, with its rural location, high-functioning residents, and homogeneous community, may not need additional safety features or community accessibility.

Other researchers have also used the PAF to describe housing programs and compare them with normative standards. For example, Thompson and Swisher (1983) used the PAF to assess the independent apartment portion of a new residential life-care facility. Safety features, space, and staff facilities were well above average, but there were relatively few orientational or prosthetic aids; at least in the apartment section of the facility, supportive features were deemphasized. Although residents saw the facility as comfortable, as reflected in their responses to the Sheltered Care Environment Scale (SCES), staff saw physical comfort as well below average. Thus, the PAF suggested that this facility was relatively well implemented with respect to its physical features; however, the SCES raised some questions about differential resident and staff reactions to this new facility.

Physical Resources in British Residential Care Homes and Hospital Facilities

Willcocks and her colleagues (1987) adapted the PAF and used it to assess 100 public residential homes for older adults in Great Britain. Although the stated goal of these homes was to give residents care and protection while allowing residents to retain control of their private world, in reality, residential life more often appeared to be a battle between individual and organizational needs. Overall, the homes' highest scores were on Safety Features and Staff Facilities; the authors viewed these high scores as reflecting the formal, institutional nature of the homes and the consequent emphasis on reduction of risk and on concrete, material manifestations of care. The homes had many physical amenities but somewhat lower scores in areas that might enhance the residents' lifestyle, such as social–recreational aids, prosthetic aids, and space availability.

The availability of physical features was related to how the facilities were organized. Homes run on a communal model had about-average resources. In contrast, semigroup homes (homes in which some activities were communal and others were organized around subgroups of residents) provided a more prosthetic environment, which was secure and negotiable; they had average or above-average scores on all aspects of the physical environment. These benefits were thought to stem from the need to adapt these buildings to the requirements of group living and from the staff enthusiasm and innovation stimulated by adoption of a new model of care. As we found in our sample of facilities in the United States, Willcocks and her colleagues (1987) found that older British homes had fewer prosthetic and orientational aids, safety features, and staff facilities.

Perkins, King, and Hollyman (1989) compared the physical features in a sample of small, privately run British residential care homes with those Willcocks and her colleagues (1987) found in their sample of public homes. As in the United States, these private residential care facilities serve many older persons with chronic psychiatric problems. The private facilities scored as high or higher on physical amenities and safety features; however, they provided fewer prosthetic, orientational, and social–recreational aids than did the public homes. The relative lack of these supportive physical features may reflect the more homelike style of the private homes, which thus may not be suitable for physically impaired residents.

Benjamin and Spector (1990) extended this work by comparing several units that serve older residents with dementia. They found that units in a residential care home tended to have more amenities and safety features than did hospital units. Similarly, Philp et al. (1989, 1991) found differences in the distribution of physical features in continuing hospital care units and community settings.

Physical Resources in Canadian Apartments

Hodge (1984, 1987) used adaptations of several of the PAF subscales to assess senior citizen apartment projects in nine small Canadian towns. Overall, only two of the projects had more than half of the physical features included in the revised checklist. More than half the projects lacked important physical features, such as call buttons in bathrooms, nonskid surfaces on stairs, handrails in the halls, bathroom doors that open out, and a grab bar at the toilet.

These findings point to a lack of adequate design implementation in these apartment projects and contrast with the sample of congregate apartments in the United States, which had an average of 71% of the PAF safety features and more than 60% of the physical amenities and social–recreational aids (Figure 3.1).

Hodge used these findings to describe specific shortcomings in the projects, especially the lack of relatively inexpensive safety features and prosthetic aids. He also recommended design standards to help develop apartment projects that would fulfill their elderly residents' needs. (For additional research using adapted versions of the PAF, see Fernandez-Ballesteros et al., 1986, 1987; Izal, 1992; Peters & Boerma, 1983; Svensson, 1984.)

Developing Program Models

Weissert and his colleagues (1989) adapted the PAF to help define distinct models of adult day care based on case mix and program affiliation. They identified three ownership/case-mix models: nursing home/rehabilitation hospital, general hospital/social service agency, and special-purpose models composed primarily of centers for veterans and for mentally ill clients. Nursing home/rehabilitation hospital centers were most likely to have physical therapy, bathing, and hair care facilities; providing these resources is consistent with their location in inpatient settings. Otherwise, however, the authors found few differences among the three models' physical facilities and equipment, despite differences in the client populations being served.

Adaptations to Other Programs

Rovins (1990) used the PAF to focus on program implementation in two residential facilities for developmentally disabled adults. A facility for young adults was more homelike with respect to community accessibility, bedroom design, and bathroom design. A facility for older adults had many prosthetic aids, but lacked personalization. Rovins pointed out that older residents often require supportive physical features that may make a facility seem more institutional. The key challenge is how to develop programs that meet older adults' need for care and medical treatment and yet do not seem impersonal.

Guiding Change Efforts

As we emphasize later (chapter 12), information about the facility and how it compares with normative standards may help staff make improvements. Hatcher and her coworkers (1983) used the PAF to focus on the quality of life of deinstitutionalized psychiatric clients living in two group homes. On the basis of the initial assessment, consultants identified the need for more features to help residents orient themselves to time and place. Accordingly, the group home manager added a few inexpensive items, such as a clock, a bulletin board, and a city map. This study illustrates how the PAF can identify simple, practical ways to improve residential facilities.

In a somewhat broader approach, Fernandez-Ballesteros and her colleagues (1986, 1987) used a Spanish version of the PAF to guide and monitor change in a large congregate apartment facility. An analysis of the physical features highlighted the need to increase prosthetic and orientational aids. Information from other parts of the MEAP pointed to the need to facilitate residents' interaction and participation in recreational activities. Accordingly, specific changes were planned in physical features (such as constructing an entrance ramp, creating new recreational areas, and rearranging furniture) and policies (such as allowing the residents to use the bulletin boards to announce activities). The changes led to increased resident participation in facility-sponsored activities, more involvement in community activities, and a greater sense of independence and influence on residents' part.

Also in this vein, Deutschman (1982) asked long-term care staff to use an adapted version of the PAF to identify key problems in their facility. In each facility, slides were taken as photographic documentation of the most frequently mentioned problems. After reviewing the slides, a team of architects, interior designers, and health care practitioners developed a variety of inexpensive, innovative options within existing codes and regulations. The information formed the content of a handbook, which can be used as a resource for staff training and problem solving.

As shown by several of the studies reviewed here, information about the physical design of residential settings can help in planning and implementing improvements. Information about how well a program is being implemented also can be obtained by considering residents' preferences, experts' judgments, or the probable impact of program features on specific groups of residents. We return to these issues in chapters 11 and 12.

RESEARCH AND POLICY APPLICATIONS

In addition to using information about the normative sample to examine program implementation, it can be used to address a number of substantive issues. For example, we have focused on the relationship between objective features of the facility and impressions formed by residents, staff, and outside observers. In order to understand more about resident selection and resource allocation, we have examined the associations between the characteristics of a facility's residents and the physical features it provides. Other researchers have used the PAF and the MEAP Rating Scale to understand the connections between physical features and resident well-being.

Perceptions of Attractiveness and Comfort

Are the objective physical features of a facility related to perceptions of its physical attractiveness and comfort? In addition to the PAF, the MEAP includes measures of how outside observers and residents view the facility's physical environment. These data make it possible to examine the relationship between physical resources and perceptions of attractiveness and comfort.

Project observers used the MEAP Rating Scale to rate the physical attractiveness and diversity of each facility (for a description of the Rating Scale, see Moos & Lemke, 1992d). Physical Attractiveness includes ratings of the facility grounds and buildings as well as of the noise level, odors, illumination, cleanliness, and condition of the facility. Environmental Diversity reflects the variety and stimulation provided by the physical environment and includes ratings of the window areas and of the variation and personalization of living spaces. In addition, a sample of residents in each facility completed the Sheltered Care Environment Scale (see chapter 5), which includes the dimension of Physical Comfort; this dimension measures the extent to which residents feel that the physical environment provides comfort, privacy, pleasant decor, and sensory satisfaction.

After controlling for level of care, we found that observers rate facilities that have more physical amenities, social–recreational aids, and space as higher on Physical Attractiveness and Environmental Diversity; residents report more Physical Comfort in such facilities (partial r's range from .19 to .53). Facilities with more social–recreational aids and safety features are also rated as more attractive by observers and as more comfortable by residents.

Thus, observers tend to rate facilities with richer physical resources as more attractive and diverse, and residents rate them as more comfortable. The findings also show that providing prosthetic and orientational aids and safety features does not necessarily reduce the perceived attractiveness and comfort.

Resident Selection and Resource Allocation

In chapter 2, we noted that the populations of residential facilities vary because of selection mechanisms, including admission policies, financing, and regulation. We have already noted evidence of some differential allocation of physical resources that results from the level of care in which the individual resides. Thus, for example, the impaired residents who live in nursing homes have more prosthetic aids available to them, and the relatively independent older people in congregate apartments have easy access to more community services.

In chapter 2, we also identified two types of resident–facility selection processes, one guided by residents' need for services, the other by predisposing and enabling factors. Need factors are the major consideration in determining the level of care. As a consequence, they help to determine the number and diversity of available physical features. However, once level of care is considered, predisposing and enabling factors are important determinants of the resources available to an individual within a given type of facility. To focus on resource allocation within a given type of facility, we conducted hierarchical regression analyses to examine the associations between physical features and residents' level of impairment, demographic characteristics, and social resources.

Need Factors

After level of care is controlled, settings with residents who are more impaired (as measured by the RESIF's Functional Abilities dimension) tend to have slightly more prosthetic features (partial r = .19). However, these facilities are less spacious and have fewer social–recreational aids (partial r's = −.26 and −.25). Such resident populations may be perceived as needing less space and fewer social–recreational aids but more prosthetic aids; these perceptions may, in turn, influence facility design, admission policies, and choice of facilities.

Predisposing and Enabling Factors

Average age is an important predisposing factor that helps to determine the physical resources available to residents. Facilities serving older residents tend to have more prosthetic aids, orientational aids, physical amenities, and space (partial r's controlling for level of care and residents' functional abilities ranged from .16 to .22). Thus, the age of residents and accompanying perceptions of frailty and vulnerability are more strongly related to the availability of supportive resources than are the residents' specific difficulties in daily activities.

Facilities with better educated residents and, to some extent, those with residents of higher occupational status, also tend to provide more physical resources, including physical amenities, social–recreational aids, prosthetic aids, safety features, and space (partial *r*'s range from .20 to .27). Facilities with more married residents offer more prosthetic aids; those with more women or more widowed residents provide more physical amenities. Once level of care and functional ability are controlled, the financial resources of residents, as indicated by the proportion of residents receiving public assistance, are not related to the physical resources available to them in the facility.

Thus, individuals who have more personal resources are likely to live in residential settings with more physical resources. After level of care is controlled, social status characteristics are more strongly related to the physical features available to residents than is their average functioning. These findings show that enabling factors make a difference in the resources available in old age. Aside from prosthetic aids, facilities serving residents who are more functionally impaired do not provide more physical resources to compensate for their deficits. In fact, more space and social–recreational aids are available where functional impairment is less, a finding that suggests that these resources are allocated according to a principle of similarity between supply and the individual's perceived ability to use them rather than to a principle of compensation. On the other hand, resident groups that are older tend to be in facilities that provide them a richer, more supportive physical environment.

Physical Features and Residents' Morale and Well-Being

As noted in the model presented in Chapter 1, systematic environmental description allows a researcher to explore the relationships between the physical resources available to residents and such outcomes as their morale and satisfaction. A number of researchers have used the PAF to help address such issues.

Nursing Home Quality and Resident Outcome

In conjunction with staff resources and an effective process of care, supportive physical features may bolster residents' functioning and thereby contribute to overall better outcomes. Braun (1991) used the MEAP quality indices (Lemke & Moos, 1986) and Rating Scale to assess 11 community nursing homes. Higher nursing home quality was associated with lower than expected 6-month mortality. More specifically, security (an index combining the PAF subscales of Prosthetic Aids, Orientational Aids, and Safety Features) was most closely associated with lower mortality. Braun speculates that the provision of physical features that support resident functioning may reflect a commitment to prevent disorientation, improve residents' adaptation to their physical limitations, and protect residents' welfare.

In support of this idea, Netten (1989) found that better lighting in residential homes was associated with better orientation among cognitively impaired residents. Oberle, Wry, and Paul (1988) found that the improved physical amenities and security features in a new hospital building were associated with lower anxiety and less use of medications for pain control among postoperative surgery patients.

Poor-Quality Community Housing and Distress Among Psychiatric Patients

Nelson and Earls (1986) assessed the community housing and social support needs of long-term psychiatric patients. The authors, using items adapted in part from the Rating Scale, developed a measure of patients' perceptions of the comfort and architectural quality of their community housing. For many clients, housing was a serious problem: More than 20% did not have enough privacy, adequate space, or access to a laundry room or recreation area. About 25% of the clients reported five or more housing problems, such as high noise level, odors, poor lighting, poor condition of walls and floors, and lack of adequate windows. Clients who boarded in a private home or boarding house reported twice as many housing problems as those who lived in their own home, in their parents' home, or in a supervised group home.

As part of the study, Earls and Nelson (1988) examined the relationship between the quality of community housing, clients' social resources, and their psychological well-being. As predicted, clients who experienced more housing problems and who had a small social network reported higher levels of distress. Moreover, among clients who had more housing problems, those who had a larger social network reported better well-being. This relationship did not hold among clients with few housing problems.

These findings indicate that clients experience more distress when they lack both physical and social resources. The findings also support the idea that social resources are more important for people who experience more housing-related stressors. Overall, poor quality housing is a chronic stressor that taxes clients' coping skills and interferes with their social adjustment and community integration.

Architectural Problems, Appraisal, and Residents' Dissatisfaction

The PAF has been adapted in a sample of 44 homes for older people in Japan (Kodama, 1986). Residents in eight of the homes reported on architectural problems in their facility. Depending on the residents' activity level and functional ability, residents emphasized different architectural problems. The most active residents focused more on the lack of social–recreational aids and community accessibility. In contrast, more functionally impaired residents focused more on the lack of prosthetic aids and safety features (Kodama, 1988b).

Actual physical features were not related to residents' distress or morale; however, residents who reported more architectural problems in their facility were more dissatisfied and had lower morale, even after controlling for their functional ability and activity level (Kodama, 1988a, 1988b). Consistent with our model (chapter 1), these findings point to the importance of personal factors and appraisal in moderating the influence of objective environmental conditions on residents' morale.

Taken together, these findings show that the physical features in a facility are associated with residents' morale and well-being. These findings provide some evidence for person–environment matching in that more active residents are more concerned about resources to support social relationships and to provide stimulation, whereas more impaired residents focus more on supportive features, such as prosthetic aids and safety features. In the next chapter, we turn to policies and services, another important set of facility characteristics that influence residents' adaptation.

4

Policies and Services

A major rationale for group residential facilities is that they allow for efficient provision of services by concentrating the recipients in one place. Once the decision is made to provide residentially based services, a number of questions arise. Should facility policies focus on selecting and maintaining a resident population that is relatively uniform in functional ability and service needs? How much does a move to a congregate setting constrain residents' autonomy? The resolution of these and other issues contributes to the quality of life that older people experience when living in a congregate residence.

As noted in chapter 1, a matching model currently underlies the financing, licensing, and regulation of facilities in the United States. This model assumes that there should be distinct facility types defined by their target population and by the mix of services available. In this model, most services are offered uniformly to all residents, with some specialized services contracted for separately or provided on an individual basis. According to this model, if a person's needs change significantly, then the person's residence should also change.

Decisions about the standards for admission, standards for continued residence, and services to be provided have indirect as well as direct influence on quality of life for residents. Whereas these decisions directly determine the characteristics of the target population and the level of services, the indirect impact is made through rules and policies governing the group. In turn, the differences in policies—such as the provision of privacy, amount of control residents have in their daily life, and flexibility of program policies—affect the quality of life for elderly residents (Bennett, 1980; Lawton, 1985b; Rodin, Timko, & Harris, 1985).

Researchers have taken several approaches to measuring and examining the impact of the policies of residential programs. For example, some investigators have used participant–observers' descriptions of nursing homes and congregate apartments to portray effectively the complex formal and informal organization of such settings (Hochschild, 1973; Ross, 1977; Schmidt, 1990; Tisdale, 1987). Other researchers have used quantifiable indices, such as size or staff–resident ratio, as measures of a setting's structural characteristics (Caswell & Cleverley, 1983; Greene & Monahan, 1981; Ullmann, 1981). In a similar vein, complex organizational features, such as the level of constraint, have been indexed by global differences, such as the type of facility or level of care provided. Thus, descriptions of the policies and services of group living settings tend to be either comprehensive and qualitative or narrowly focused and quantitative.

There are some notable exceptions to this general rule. For example, Lieberman and Tobin (1983) contrasted policies and services that characterize institutional versus community life. In the study, a judge used observational data and interviews with the administrator to rate 26 facilities; the ratings were used to identify 11 dimensions indicative of the quality of a facility. These dimensions measure the extent to which residents are encouraged to engage in goal-directed activities; how much residents are treated as individuals and recognized for their accomplishments; the level of tolerance for deviant behavior; the adequacy of health care; and the amount of warmth, stimulation, and social integration in the facility.

Building on these ideas, Booth (1985) developed an Institutional Regimes Questionnaire to assess four aspects of the policies in British homes for older people: personal choice, privacy, freedom and independence, and resident participation in facility governance. These dimensions differentiated somewhat among the homes, but the differences were not related to changes in residents' levels of dependency or to mortality rates.

In an effort to develop a measure of facility policies and services that is both comprehensive and quantitative, we tried to extend and integrate these approaches. We wanted the method to provide detailed information that could enrich descriptive studies of individual facilities and to be comprehensive enough to apply to surveys of a broad range of facilities. The result of our efforts is the Policy and Program Information Form (POLIF).

ASSESSING POLICY AND PROGRAM RESOURCES

The Policy and Program Information Form (POLIF)

The POLIF measures the policies and services of group living facilities. In general, the information is obtained from the facility administrator, but staff members and residents can serve as alternate or additional sources of information. The POLIF has 130 items and taps nine aspects of the policies and services in a facility (Table 4.1).

The nine dimensions fall into three groups. The first two dimensions reflect the extent to which behavioral requirements are imposed on residents. Expectations for Functioning measures the minimum acceptable capacity to perform daily living tasks. Acceptance of Problem Behavior assesses the extent to which the facility accepts uncooperative, aggressive, or eccentric behavior.

The second set of POLIF dimensions measures the balance between individual freedom and institutional order and stability. The four areas

Table 4.1 Policy and Program Information Form (POLIF) subscale descriptions and item examples

BEHAVIORAL REQUIREMENTS FOR RESIDENTS	
1. Expectations for Functioning (11 items)	Minimum capacity to perform daily living functions that is acceptable in the facility (Is inability to clean one's own room tolerated? Is incontinence tolerated?)
2. Acceptance of Problem Behavior (16 items)	Extent to which aggressive, defiant, destructive, or eccentric behavior is accepted (Is refusing to bathe tolerated? Is pilfering or stealing tolerated?)
INDIVIDUAL FREEDOM AND INSTITUTIONAL ORDER	
3. Policy Choice (19 items)	Extent to which the facility allows residents to individualize their routines (Is there a curfew? Are residents allowed to drink a glass of wine or beer at meals?)
4. Resident Control (29 items)	How much residents are involved in facility administration and influence facility policies (Is there a residents' council? Are residents involved in deciding what kinds of new activities or programs will occur?)
5. Policy Clarity (10 items)	Extent of formal institutional mechanisms for defining expected behavior and communicating ideas (Is there a handbook for residents? Is there a newsletter?)
6. Provision for Privacy (10 items)	Amount of privacy given to residents (How many residents have private rooms? Are residents allowed to lock the doors to their rooms?)
PROVISION OF SERVICES AND ACTIVITIES	
7. Availability of Health Services (8 items)	Availability of health services in the facility (Is there an on-site medical clinic? Is there physical therapy?)
8. Availability of Daily Living Assistance (14 items)	Availability of facility services that assist residents in tasks of daily living (Is there assistance with personal grooming? Is dinner served each day?)
9. Availability of Social–Recreational Activities (13 items)	Availability of organized activities within the facility (How often is there outside entertainment? How often are there classes or lectures? How often are there parties?)

cover the latitude of choice residents have to establish their own daily routine (Policy Choice), formal institutional structures giving residents potential influence in the facility (Resident Control), formal arrangements for communicating about policies and programs (Policy Clarity), and privacy available to residents (Provision for Privacy).

The third set of POLIF dimensions measures the availability of services and activities. The dimensions tap health services, daily living assistance, and social–recreational activities. (For more information about the normative and psychometric characteristics of the POLIF, see Moos & Lemke, 1992c; for a description of the initial version of the POLIF, see Lemke & Moos, 1980.)

Policies and Services in Different Types of Facilities

Figure 4.1 shows the means of the nine POLIF dimensions for the 135 nursing homes, 60 residential care facilities, and 67 apartments in the community sample (results for the veterans sample are presented in chap-

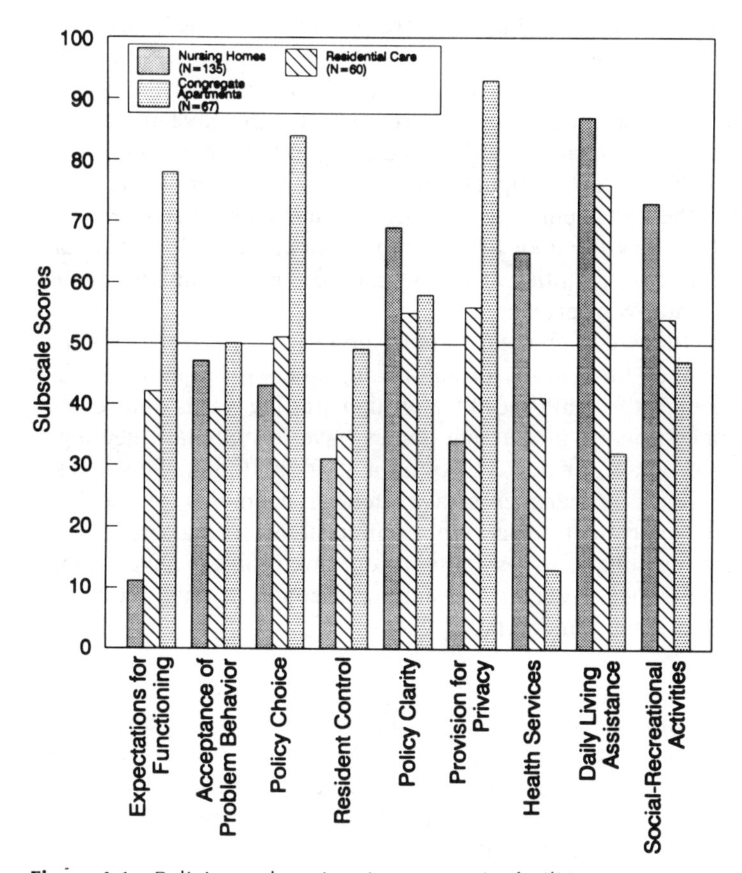

Figure 4.1 Policies and services in community facilities.

ter 6). The means reflect the percentage of applicable items in the scored direction. For example, Expectations for Functioning comprises 11 questions about policies concerning resident disability and dependence. Facility scores reflect the percentage of these items on which the facility policies discourage disability or dependence. On average, the 135 nursing homes set requirements in just 11% of these areas of functioning.

Acceptance of Problem Behavior is composed of 16 items; facility scores can vary from 0% (no tolerance of nonconformity) to 100% (tolerance of all 16 aspects). If a facility tolerates nonconformity for 8 of the 16 items, its score is 50%. On average, the nursing homes accept problem behavior for almost half (47%) of the examples included in the POLIF. Similarly, on average, the nursing homes provide residents with choice for 43% of the Policy Choice items, with control for 31% of the Resident Control items, and so on.

Policies

As with the PAF, the norms for the three sets of community facilities reflect meaningful variations between types of settings. On most dimensions, the average scores parallel the level of care. As expected, nursing homes have much lower expectations for resident functioning than do residential care facilities, which are lower on this dimension than are congregate apartments. Similarly, nursing homes are much lower than apartments on policy choice and resident control. Apartments have much more privacy than residential care facilities, and they, in turn, have more than nursing homes. Nearly all the apartments provide private rooms and private baths for their residents, give residents their own mailboxes, and allow residents to lock the door to their rooms. In contrast, less than half of the residential care facilities and less than 10% of the nursing homes provide these features of privacy.

Of the three types of facilities, nursing homes are highest on policy clarity; that is, they have more formal arrangements for communicating policies, particularly to staff. Because of high staffing levels and regulatory requirements, nearly all nursing homes have formal staff meetings, an orientation program for new staff, and a staff handbook. Fewer residential care facilities or apartments have such arrangements for staff. On the other hand, apartments place more emphasis on formal procedures for communicating policies to residents. About three-fourths have a handbook for residents and a regular newsletter; in contrast, only about half the residential care and nursing facilities do.

Services

The scores for the service-availability dimensions closely parallel the level of care. As expected, nursing homes provide more health services, a greater variety of assistance in daily activities, and more social activities than do residential care facilities. In turn, residential care facilities provide more of these services than do apartments.

Overall, the major distinctions between types of community facilities are in their expectations of resident functioning, the services they provide, and the degree of institutional constraint they impose on residents. This pattern reflects the application of a matching model in which facility types have distinctive service packages and policies regarding acceptable resident functioning; such policies have consequences for resident autonomy.

Nursing homes are tolerant of disabilities and have a variety of support services but are lowest on the dimensions related to resident autonomy; such policies reduce freedom. Residential care facilities, which are intermediate in expectations for resident functioning and in making support services available, are like nursing homes in the level of resident involvement in decision making, but they offer slightly more policy choice and more privacy than do nursing homes. Apartments, which expect residents to be relatively free of impairment, offer few supportive services but provide residents the most opportunities to exercise control over their use of time, personal objects, space, and facility policies.

Specific Policies and Services in Different Types of Facilities

As with physical and architectural features, we can examine the presence of specific policies and services in the three groups of community facilities. The findings help to clarify the range of facilities available to older people and provide a more detailed picture of how the matching model is implemented in these facilities.

Expectations for Functioning and Acceptance of Problem Behavior

Gerontological experts and social policy goals suggest that a spectrum of group residential facilities can best provide options for a diverse population of older people (Brody, 1982). These options should be broad enough to serve older people who are quite dependent in their day-to-day functioning as well as those who are functioning independently but need resources to compensate for losses or lack of social supports. We have noted the controversy over the extent to which residents should be segregated into different facilities according to their level of functioning and type of disability.

The POLIF dimensions of Expectations for Functioning and Acceptance of Problem Behavior measure the range of residents' functioning and competence that facilities will accept. As shown in Figure 4.1, congregate apartments have the highest standards for resident functioning, followed by residential care facilities and nursing homes. Nearly all apartments expect residents to be continent, to feed and bathe themselves, and to be cognitively competent (Table 4.2). About two-thirds of apartments also require residents to be ambulatory and to be able to clean their own rooms. In contrast, only a few of the nursing homes have requirements in any of these areas. Most residential care facilities require residents to be

Table 4.2 Percent of facilities responding in the scored direction on selected Expectations for Functioning and Acceptance of Problem Behavior items

Item	Percent in scored direction		
	Nursing homes ($N = 135$)	Residential care facilities ($N = 60$)	Congregate apartment facilities ($N = 67$)
ARE THE FOLLOWING BEHAVIORS DISCOURAGED OR INTOLERABLE?			
1. Incontinence	9	81	97
2. Inability to feed oneself	8	73	97
3. Inability to walk	3	85	71
4. Inability to bathe or clean oneself	7	32	94
5. Confusion or disorientation	9	36	82
6. Inability to clean one's room	7	17	67
7. Depression (frequent crying, sadness)	21	44	68
ARE THE FOLLOWING BEHAVIORS TOLERATED?			
1. Refusing to take prescribed medicine	20	33	70
2. Being drunk	24	36	58
3. Leaving the building in the evening without letting anyone know	37	73	100
4. Creating a disturbance; being noisy or boisterous	87	63	48
5. Verbally threatening another resident	66	42	31
6. Damaging or destroying property, e.g., tearing books or magazines	58	32	27
7. Physically attacking a staff member	37	14	9
8. Physically attacking another resident	28	5	4

The three sets of facilities differ significantly ($p < .01$) on all items.

ambulatory, continent, and able to feed themselves but permit dependence in housekeeping activities. In other areas of functioning, residential care facilities vary widely in their policies, which reflects the heterogeneity of this category of facilities.

Apartments, which are most accepting of the independent status of residents, are also most likely to accept noncompliant behavior that affects mainly the individual involved; such behaviors include refusing to take prescribed medications, being drunk, or wandering around the buildings or grounds at night. In contrast, nursing homes and residential care facilities emphasize communal living, assume responsibility for medications, and place a premium on reducing risk, and accordingly are less likely to accept these problem behaviors. However, of the three types of facilities, nursing homes are most likely to accept problem behaviors that may arise from mental confusion, such as being noisy or boisterous, verbally threatening another resident, or damaging property.

Residential care facilities, which accept residents with a moderate level of disability, are less tolerant of individual nonconformity than are apartments and less tolerant of aggressive or destructive behavior than are nursing homes; as a consequence, these facilities are least accepting of problem behaviors. Consistent with these results, Mor, Sherwood, and Gutkin (1984) found that residential care managers were more likely to accept a physically impaired client than a client with behavioral or emotional problems.

Viewed from another perspective, older people who need some help with housekeeping can find a place in most residential care facilities or in some apartments; options are somewhat more limited for those who also are depressed or confused. Nursing homes and some residential care facilities are available to older people who are dependent in self-care activities, such as dressing and bathing, but being nonambulatory or incontinent almost completely restricts one's choice to a nursing home, as do behavior problems such as aggressiveness. Older people with these problems are likely to be most difficult to place, even in a nursing home. At the policy level, at least, the three types of facilities overlap little in their target populations.

These findings highlight the limited residential options for physically able but emotionally or behaviorally disturbed older people; providers' policies often preclude placement in residential care or apartment facilities. Instead, these older people often are placed in nursing homes where staff rely on psychotropic medications and physical restraints to manage their behavior problems. Mor and his colleagues (1984) point to the need for provider education and financial incentives in residential care facilities to counteract providers' bias against people with mental or behavior problems.

In their survey of residential care facility managers, Mor and his colleagues found that actual admission practices were even more skewed against those with behavioral or emotional problems than were stated policies. Similarly, our data allow us to examine the match between stated policies and resident characteristics. We found that for feeding, dressing, walking, and bathing, about 85% of all facilities housed a resident population consistent with their stated policies. Approximately 7% had policies that were more liberal than their actual practice; that is, their policies permitted resident dependence in these activities, but the facilities reported that all current residents were independent in these areas. A similar proportion reported that impairment in these areas would not be accepted yet had some residents who needed assistance in these activities.

Table 4.3 Percent of facilities responding in the scored direction on selected Policy Choice, Resident Control, and Provision for Privacy items

Item (scored direction in parentheses)	Percent in scored direction: Nursing homes (N = 135)	Residential care facilities (N = 60)	Congregate apartment facilities (N = 67)
POLICY CHOICE			
1. Is there a set time at which residents are awakened in the morning? (no)	30	36	98
2. Is there a set time for baths? (no)	24	51	100
3. Is there a set time at which residents are expected to go to bed? (no)	66	59	98
4. Are residents allowed to skip breakfast to sleep late? (yes)	31	59	95
5. Can residents choose to sit wherever they want at meals? (yes)	74	42	88
6. Is there an hour's range or more during which residents can choose to eat lunch? (yes)	25	25	60
7. Are residents allowed to have alcohol in their rooms? (yes)	28	34	75
8. Are residents allowed a hot plate or coffeemaker in their room? (yes)	3	22	85
9. Are residents allowed to move furniture in the room? (yes)	60	76	99
RESIDENT CONTROL			
1. Are there regular house meetings for all residents? (yes)	48	46	82
2. Is there a residents' council? (yes)	64	39	73
3. Are there committees that include residents as members? (yes)	26	25	63
4. Do any of the residents perform chores or duties in the facility? (yes)	45	69	69

(continued)

Table 4.3 *(continued)*

Item (scored direction in parentheses)	Percent in scored direction: Nursing homes ($N = 135$)	Residential care facilities ($N = 60$)	Congregate apartment facilities ($N = 67$)
5. Do residents help decide what new activities or programs will occur? (yes)	14	31	49
6. Do residents help plan welcoming or orientation activities? (yes)	9	21	40
*7. Do residents have input in planning menus? (yes)	45	55	49

The three sets of facilities differ significantly ($p < .01$) on all items except the one marked with an asterisk (*).

Overall, our results show a relatively high consistency between facility policies and the actual characteristics of their resident populations.

Policy Choice and Resident Control

Two of the most salient facility characteristics are the flexibility of scheduling and procedures and the availability of opportunities for residents to participate in facility governance. As shown in Figure 4.1, apartments offer their residents much more choice and control than do residential care or nursing facilities. Consequently, impaired residents tend to live in more structured and regimented facilities.

This pattern is further reflected in Table 4.3, which shows that apartments provide residents with the most options for individual patterns of daily life, and nursing homes provide the fewest. Residential care facilities are only slightly more flexible than are nursing homes.

Apartment facilities also allow residents more independence in their own rooms than do residential care or nursing facilities. For example, most apartments permit residents to have a hot plate or coffeemaker and to have alcohol in their own rooms; however, most nursing homes and residential care facilities do not. In most facilities, residents can bring their own furniture with them and rearrange the furniture in their rooms, but here again, apartments have the most liberal policies. The majority of apartments and nursing homes allow residents to drink a glass of wine or beer with their meals and to choose their own seats at meals; alcohol at meals and seating are more likely to be regulated by residential care facilities.

With respect to Resident Control, apartment facilities are more likely than either nursing homes or residential care facilities to delegate responsibilities to their residents. The majority of apartments schedule regular house meetings, have a residents' council and committees that include residents, give residents chores within the facility, and involve residents in planning the activity program (Table 4.3). Residential care facilities are as likely to give residents chores, and nursing homes are about as likely to have a residents' council, but in general, fewer of the nursing homes or residential care facilities provide resident control in these areas. Apartment facilities are also more likely to involve residents in orientation programs for new residents.

These findings show that apartments tend to give their residents flexibility in scheduling activities and using their rooms and to provide residents means for voicing their opinions and influencing decision making. Most of the day-to-day administration remains in the hands of staff. In contrast, residents in nursing homes have given up much of their control over daily routines and use of personal space. Their lives tend to be regimented to facilitate the care routines of staff; their personal control is limited by the absence of personal space, the intrusiveness of communal life, and the concerns about risk reduction. Although older people in residential care facilities tend to have only moderate functional impairment, their lives are nearly as regimented as those of nursing home residents, and they have little opportunity for a formal voice in the facility.

Health Services and Daily Living Assistance

Just as the data on policies indicate that different facility types focus on different target populations, the data on services confirm that most congregate settings are administered according to a matching model: Different facility types offer service packages intended to match the needs of their residents; when residents' functioning changes, they are moved to different facilities.

Only nursing homes explicitly accommodate residents with major health problems, as indicated by policies accepting incontinence or the need for assistance in eating or in transferring from bed to chair. To meet the needs of these very impaired residents, all the nursing homes have regularly scheduled nurses' hours and supervise medications; most provide some physical therapy (Table 4.4). To meet the daily living needs of very impaired residents, all the nursing homes provide housekeeping and assistance with personal care and grooming. Most also assist residents with shopping and banking and provide access to religious services.

Even though most nursing homes accept residents who are depressed, confused, or withdrawn, only about half offer any form of psychological counseling or therapy. Similarly, although almost all nursing homes accept nonambulatory residents, only about two-thirds of the facilities provide transportation for residents. Thus, nursing homes meet their target population's medical care and supervision needs, but some facilities do not adequately address the needs of residents with emotional or behav-

Table 4.4 Percent of facilities providing selected health services, daily living assistance, and social–recreational activities

Item	Percent in scored direction: Nursing homes ($N = 135$)	Residential care facilities ($N = 60$)	Congregate apartment facilities ($N = 67$)
HEALTH SERVICES			
1. Regularly scheduled nurses' hours	100	43	36
2. Assistance in using prescribed medications	100	93	14
3. Physical therapy	85	27	6
4. Psychotherapy or personal counseling	54	37	26
DAILY LIVING ASSISTANCE			
1. Assistance with housekeeping or cleaning	100	95	35
2. Lunch at least 5 days a week	100	100	68
3. Assistance with personal care or grooming	100	88	14
4. Assistance with banking or other financial matters	83	57	26
5. Barber or beauty service	99	82	48
6. Assistance with shopping	89	72	45
SOCIAL–RECREATIONAL ACTIVITIES			
1. Religious services	100	86	67
2. Arts and crafts	95	61	86
3. Exercises or physical fitness activity	96	68	69
4. Movies	90	61	55
5. Discussion groups	93	63	63
6. Self-help or mutual support group	59	27	25
*7. Classes or lectures	55	42	64

The three sets of facilities differ significantly ($p < .01$) on all items except the one marked with an asterisk (*).

ioral problems or of residents whose mobility impairments isolate them from the community.

Most residential care facilities will accept residents who have difficulty with housekeeping tasks or personal grooming and bathing, but few have policies to accommodate residents with impairments such as incontinence or being in a wheelchair. Consistent with these policies, residential care facilities are almost as likely as nursing homes to offer basic assistance with daily living tasks. Thus, frail and moderately impaired older people can obtain most of the necessary support services for their daily living needs in a residential care facility, but therapeutic services are rare, and residents with health care needs have to obtain these services elsewhere.

Only about one-third of the congregate apartments have policies allowing those with mobility impairment, depressed mood, or moderate physical disabilities to reside there; their services match relatively well an intact target population.

These results indicate that, as with policies, facility types overlap little in the service packages they provide. Generally speaking, provision of services is guided by a model in which different facility types offer a mix of services aimed at distinct subgroups of the older population. Facility policies on resident disability appear relatively well matched with the services provided, with the exception that residents' mental health needs and need for community contact are poorly addressed in nursing homes, and residents' needs for rehabilitative services are poorly met in residential care facilities.

Social–Recreational Activities

Nursing homes have somewhat more social–recreational activities than residential care facilities, which have somewhat more than do congregate apartments. Nearly all nursing homes offer religious services, arts and crafts, and exercise or other physical activity at least once a month (Table 4.4). Nursing homes also are most likely to show movies and to organize discussion groups. Nursing homes tend to use the activity program to meet some of the mental health needs of their residents; over half have self-help or mutual support groups. A majority also offer classes or lectures for residents at least once a month. Thus, the activity program in nursing homes may be used to compensate for limited professional and rehabilitative services, such as occupational or physical therapy or counseling services. The activity program in residential care and apartment facilities focuses on entertainment; it provides few activities with an explicitly therapeutic focus.

Nursing homes are designed to provide a self-contained environment that allows residents to remain in the facility and have most of their entertainment and service needs met; the costs come in terms of greater regimentation of their lives and the provision of some services that are not actually needed. Residential care facilities in large part target a population with moderate physical impairment, for which they offer most of the appropriate services; as with nursing homes, access to these services comes at the cost of some regimentation, loss of autonomy and privacy, and with little compensating control over facility management. Congregate apartments, which generally expect residents to maintain independent functioning, provide few supportive services beyond a meal plan and some organized activities; residents in these facilities retain a high level of autonomy and privacy and are given some voice in facility governance.

POLICIES, SERVICES, AND THE ADEQUACY OF PROGRAM IMPLEMENTATION

As with the PAF, information from the POLIF can be depicted as a profile comparing the facility's policies and services with those of the normative sample of similar settings. Here, we present the POLIF profiles of the three facilities whose physical features were presented in chapter 3. These profiles show standard scores with a mean of 50 and standard deviation of 10. The profiles indicate how well a program is implemented from a normative perspective. They identify the areas in which a facility has more or fewer policy resources and services compared with other facilities offering a similar level of care. (For additional examples of POLIF profiles, see Lemke & Moos, 1985; Moos & Lemke, 1983, 1984, 1992c; Moos, Lemke, & Clayton, 1983; Moos, Lemke, & David, 1987.)

Middleton Nursing Home

Middleton Nursing Home, a larger than average nursing facility in a midwestern city, has a rather typical resident population for a nursing home. The one exception is the level of resident disability, which is fairly high. As we saw in chapter 3, Middleton offers residents many prosthetic and orientational aids, a variety of features to make the setting pleasant and to stimulate social activity, and access to community resources. On the other hand, it lacks safety features and offers limited space. Figure 4.2 shows the POLIF results for Middleton Nursing Home.

Typical of most nursing homes, Middleton's policies set few requirements regarding residents' ability to function independently in activities of daily living. However, problem behavior is accorded little acceptance. Whereas most nursing homes tolerate such behaviors as threatening another resident, refusing to bathe, or damaging property, Middleton residents who behave this way are asked to leave the facility.

Policies give residents a high level of choice. In contrast, residents have only about-average opportunities for input in decision making. The arrangements for communicating policies to residents and staff and the provisions for privacy are similar to those found in nursing homes generally.

Finally, Middleton is typical of nursing homes in the services and activities it provides. Physical therapy and counseling are available, along with such standard services as an on-call doctor, regular nurses' hours, and assistance with medications. The facility provides religious advice, assistance in handling spending money, and a beauty service, as well as housekeeping, personal care, and laundry service. The activity program also is representative.

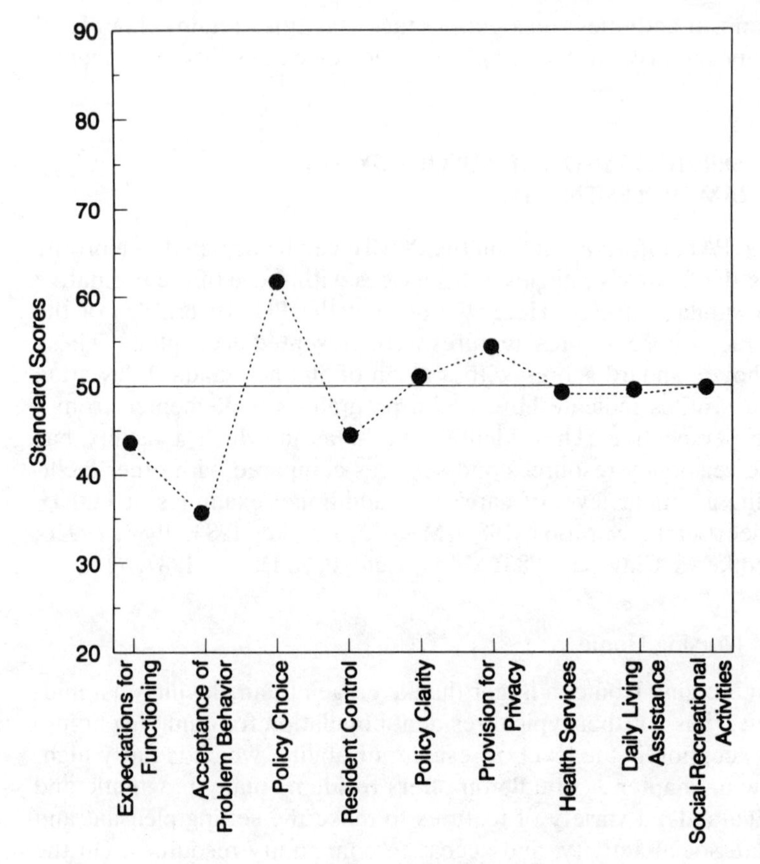

Figure 4.2 Policies and services in Middleton Nursing Home.

Maude's Community Care Home

Maude's, a small family-run home with a mostly male population, is located in a West Coast suburb. As seen in chapter 3, the emphasis is on providing a safe environment with easy access to the community.

The policies set average expectations in regard to both independence in daily activities and conformity to normative behaviors (Figure 4.3). The expectations regarding functional independence are consistent with the average level of functioning shown by residents.

Compared with other residential care facilities, Maude's provides residents a low level of choice and control. There are set times for meals, getting up in the morning, bathing, and going to bed. Although residents can drink alcohol at meals, they are not allowed to have liquor in their rooms. There are few formal arrangements for giving residents a voice in decision making. Unlike a majority of residential care facilities, there is neither a residents' council nor house meetings.

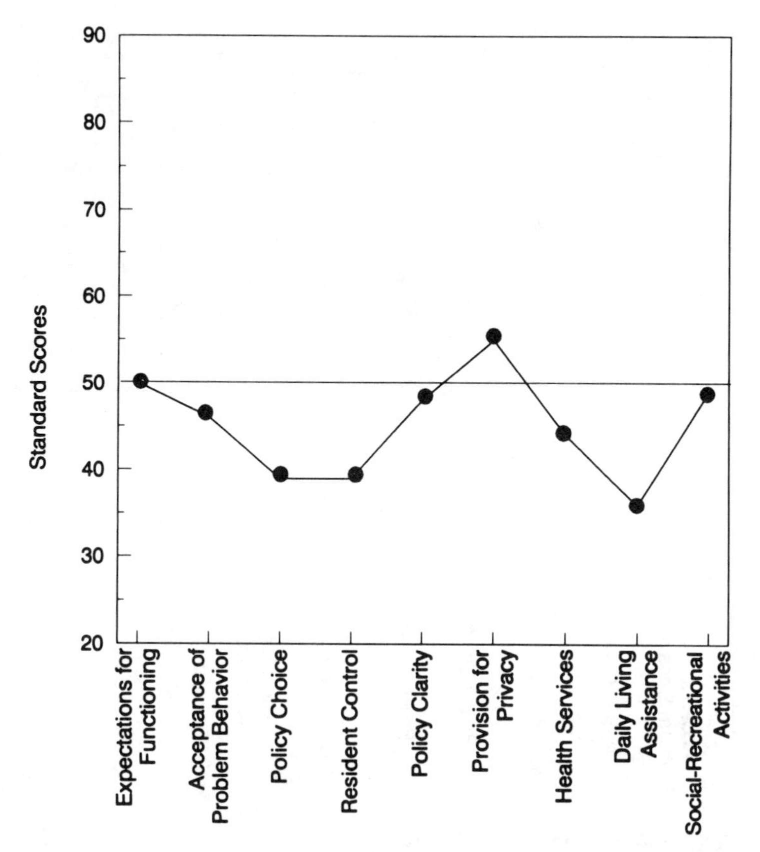

Figure 4.3 Polices and services in Maude's Community Care Home.

On the other hand, Maude's has the usual procedures for communicating policies and for providing residents with privacy. Although the facility is licensed to house 30 residents, mostly in two-person rooms, it is operating at just over half its capacity; consequently, most residents have their own room and bath and are permitted to close the door to their rooms.

Maude's provides organized activities and health services common to residential care facilities, but only basic assistance in daily living, such as housekeeping, preparation of meals, and assistance with personal care.

Amity House

Amity House is a rural apartment complex developed by a religious organization. It serves residents from relatively privileged backgrounds. Amity House is isolated from community resources but offers residents ample space and numerous amenities.

The policies regarding impairment and problem behavior are typical of apartments (Figure 4.4). The facility emphasizes giving residents a formal voice in decision making and systematically communicating policies to

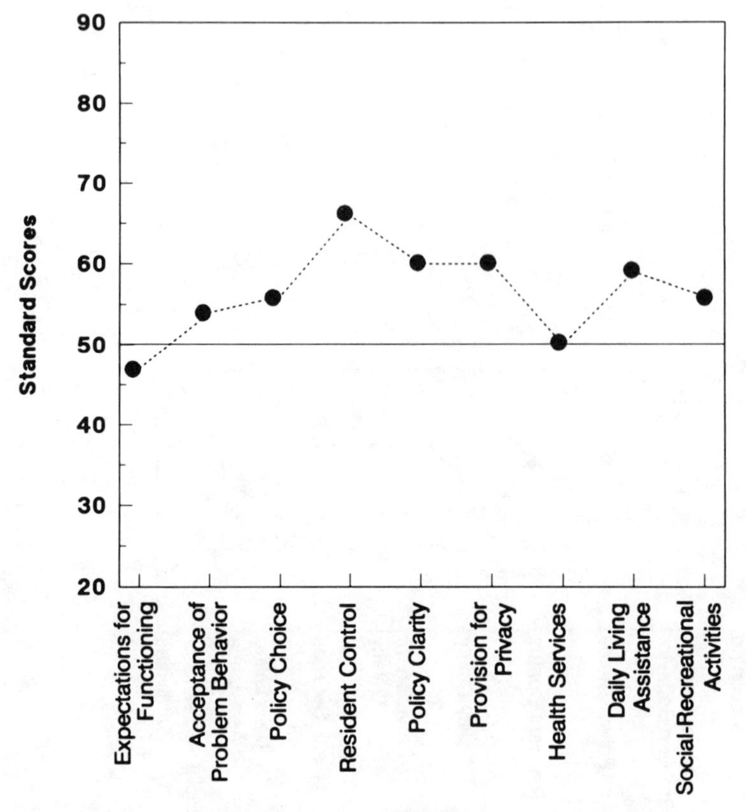

Figure 4.4 Policies and services in Amity House.

staff and residents. Residents' privacy is well provided for and residents are given about-average opportunities for choice in their daily routines.

The service emphasis for Amity House is on daily living assistance (three meals a day, housekeeping service, transportation, assistance with banking, and a barber/beauty service), with average levels of health care services and organized activities.

Program Functioning and Standards of Implementation

These three examples illustrate extensive variation in the policies and services of group residential facilities. Compared with other nursing facilities, Middleton gives residents above-average choice but has little acceptance of problem behavior. Compared with other apartments, Amity House offers a relatively involving and service-rich environment. In contrast, Maude's is average or below in all the policy and services areas except privacy.

One standard for assessing the adequacy of program implementation is the conditions in a normative sample of facilities. The POLIF profiles embody this approach by comparing each program with facilities that provide the same level of care. In a similar manner, several researchers have used the POLIF to assess program implementation in nursing homes and other long-term care facilities (Billingsley & Batterson, 1986; Cole, 1982; Johnson, 1981; Thompson & Swisher, 1983), psychiatric inpatient units (Garritson, 1986), and group homes (Manning, 1989).

Together with information about resident and staff preferences and research on program impact, this information can be used to help administrators improve their facilities. For example, Maude's is relatively low on choice and control compared with other residential facilities. Given that resident functioning at Maude's is about average and that greater choice and control tend to have positive consequences for such residents (chapter 8), the administrator might provide a somewhat more flexible daily schedule and allow residents to participate more actively in decision making. Information about residents' preferences can also help guide planning for specific changes (chapter 11).

Identifying Program Models

Program models can help gerontologists formulate standard criteria for describing and classifying long-term care settings. Such models can provide normative and conceptual standards by which to measure program implementation, specify the types of residents best managed in existing programs, and identify and plan new programs needed to provide a more integrated array of services. These models may also help to focus on the congruence between residents' medical and psychosocial needs and a program's policies and services.

To understand the underlying models that have guided different types of long-term care programs, Braun and Rose (1989) characterized nursing care programs in terms of staff expectations of resident functioning and behavior, ability to meet nursing care needs, and program goals and constraints. They compared five long-term care programs: skilled nursing facilities, intermediate care facilities, foster homes, day hospital, and comprehensive home care. The POLIF was used to measure program expectations of resident functioning and acceptance of problem behavior.

Consistent with our findings, program types differed in their admission policies and services. The foster home and day hospital programs had the highest expectations for resident functioning, followed by the intermediate care and skilled nursing facilities. For example, the foster home program discouraged admission of those unable to get out of bed and of those who were very depressed. The home care program was the most accepting of disabilities and problem behaviors.

Somewhat unexpectedly, the skilled nursing and intermediate care facilities did not tolerate problem behaviors such as residents' refusing to take prescribed medicine or refusing food or drink. The programs in which people remained in their own home (day hospital and home care) were more accepting of these problem behaviors. In general, the skilled nursing and home care programs provided the most nursing services and managed the most impaired and physically ill residents; intermediate care facilities did not meet residents' medical care needs any better than did the foster homes.

In a comparable project, Conrad, Hanrahan, and Hughes (1988) followed our conceptual framework to construct the Adult Day Care Assessment Procedure (ADCAP). Developed on data obtained from more than 900 adult day-care programs in the United States, ADCAP measures characteristics of the client population, services and activities provided, and the quality of the social environment.

Conrad and his coworkers used information from the ADCAP to define an Alzheimer's day-care program model; they compared programs in which at least 70% of the clients had Alzheimer's disease and programs in which none of the clients had Alzheimer's disease. The two sets of programs provided services and activities that were generally consistent with their clients' levels of impairment. Compared with non-Alzheimer's programs, Alzheimer's programs provided more case management, health evaluation, training in activities of daily living, therapeutic activity, and family involvement.

Normalization in Community Care Homes

Planners and practitioners expected the process of deinstitutionalizing chronically mentally ill and frail older people to provide an environment that promotes self-reliance and empowers residents as consumers of care. To examine whether this type of environment has been implemented, Blake (1985–86) used the POLIF and other parts of the MEAP to assess the extent of normalization in a sample of 10 New Jersey boarding homes.

In contrast with social policy goals, the homes had relatively restrictive policies and offered few therapeutic services or social–recreational activities. Expectations for functioning in daily living activities were moderate, but residents generally had few opportunities to perform chores and fulfill normal responsibilities in the homes. Moreover, even though the homes had little tolerance for problem behavior, they made few efforts to convey these rules clearly.

Residents had little voice in the management of their homes; although a residents' council helps to empower residents, none of the homes had such a governance mechanism. Blake concluded that these boarding homes were oriented toward the needs of staff rather than toward the provision of treatment or rehabilitation. Blake's study shows that community-based residential settings can be as restrictive as hospital-based or nursing home settings, a conclusion that is consistent with our findings in psychiatric settings (Moos, 1988b).

In contrast, some findings from the United Kingdom suggest that small residential homes may develop normative resident-oriented environments. Perkins and her colleagues (1989) found that private homes for older adults with chronic psychiatric problems provide their residents with a high level of privacy, autonomy, and choice. According to Benjamin and Spector (1990) and Philp and his coworkers (1989), British residential care facilities provide more privacy, choice, and resident control than do long-term care hospital facilities. Sewell and Lethaby (1990) obtained similar findings in a sample of New Zealand residential and rest homes. As expected, older adults tend to be more satisfied and to function better in residential homes than in long-term hospital units (Perkins et al., 1989).

When clients' families play an advocacy role, they may be able to improve program implementation and the quality of group homes. Wilson and Kouzi (1990) compared group homes for mentally disabled clients with group homes for developmentally disabled clients. Facilities for developmentally disabled clients offered more personal care and daily living assistance services. These facilities were also smaller and had better physical features and better-trained staff. The authors speculated that the higher quality of group homes for developmentally disabled clients was due to the intensive advocacy efforts exerted by their family members.

RESEARCH AND POLICY APPLICATIONS

As with the PAF, the normative data we have obtained with the POLIF permit us to address a number of substantive issues. Chapters 8 and 9 focus on the issues of choice and control. Here we examine evidence on resident selection and resource allocation. In addition, we summarize the work of other researchers who have used the POLIF or measures adapted from it to address important research and policy issues, in particular the links between facilities' policies and residents' morale and well-being.

Resident Selection and Resource Allocation

We have seen evidence for selection mechanisms that operate to channel residents into facilities that offer different levels of care. As a consequence, the more impaired residents who live in nursing homes have access to more services but are given fewer opportunities for choice, control, and privacy. The moderately impaired residents in residential care facilities have somewhat higher levels of choice and privacy and lower levels of services. Apartment residents have access to fewer services but exercise more choice and control and enjoy more privacy. Here we focus on whether similar processes operate within levels of care.

We examined the relationships between the facility's policies and services and the same sets of resident characteristics considered in relation to physical features in chapter 3: need factors, as indexed by residents' functional abilities, and predisposing and enabling factors, as reflected primarily by residents' demographic and social resources. As before, we conducted hierarchical regression analyses to predict facility policies and services; we entered level of care, residents' functional abilities, and then residents' demographic characteristics and social resources.

Need Factors

If the matching model is being applied within facility types, we would expect to find a positive relationship between residents' need for services and the services provided them; this relationship would extend beyond the gross differences distinguishing types of facilities. In fact, after level of care is controlled, the more impaired residents are in facilities with lower expectations for functioning. Thus, within each facility type, facility policies are generally consistent with the actual characteristics of the resident population. As might be expected from the findings for facility types, facilities with more-impaired residents offer less resident control and privacy. Contrary to expectation, however, facilities with more-impaired residents provide fewer daily living services and no more health services than do facilities with higher-functioning residents.

Predisposing and Enabling Factors

After level of care and residents' functional abilities are controlled, expectations for functioning and acceptance of problem behavior generally are not associated with residents' age or social resources. However, facilities whose residents achieved higher educational and occupational status tend to provide more privacy and social–recreational activities. Facilities with more married residents offer higher levels of policy clarity, choice, control, and privacy. Similarly, facilities with more women residents emphasize choice, privacy, and a rich program of formal activities. Finally, privacy and the number of organized activities vary with the average age of residents. Facilities serving older populations offer residents more privacy as well as more organized activities, perhaps as a response to perceived withdrawal of interest or decrements in residents' abilities to arrange their own leisure activities.

The earlier findings (chapters 2 and 3) showed that residents with more social resources tend to be in better-staffed facilities that have more stimulating and supportive physical features. The findings presented here show that these advantages extend to more flexible policies and richer services. These relationships are likely to reflect a mutually reinforcing pattern in which residents with higher status have more leeway in facility selection, place more emphasis on flexible policies and richer services in selecting a facility, and support these aspects of facility management once they become residents. Furthermore, social background and expectations appear to have a stronger influence than economic resources in determining the facility's provision of these opportunities: Educational, occupational, and marital status, rather than the number of private paying residents, is related to policies that are more liberal and to services that are more diverse.

Policies, Morale, and Well-Being

The policies in congregate settings are of interest in and of themselves; researchers have also focused on their impact on such resident outcomes as morale, well-being, and functional independence (e.g., Kruzich, Clinton, & Kelber, 1992).

Residential Care Homes

Willcocks and her colleagues (1987) assessed the policy emphasis in British residential care homes and examined its relationship to resident and staff morale. They adapted items from the POLIF to describe how much a residential home is oriented toward resident versus staff needs. Resident-oriented policies were indicated by high resident choice and control; staff-oriented policies were indicated by lack of resident involvement in facility organization and governance.

As predicted, Willcocks and her coworkers found that in homes populated by residents with more physical impairments, policies were more oriented toward staff needs. In contrast, homes that had fewer mentally infirm residents tended to have more resident-oriented environments and to offer more opportunity for stimulation and independence. In homes with a higher staff-to-resident ratio, staff were more likely to encourage resident autonomy. Although we found no such relationship in our data, we did find that facilities with better-trained and more-experienced staff had clearer policies that were more oriented toward residents' needs. According to Willcocks and her colleagues, staff may pay a price for policies that focus too much on resident needs: Staff reported more worries and anxiety in homes that had more resident-oriented policies.

Work by other researchers suggests that there need not be a direct trade-off between resident and staff control. Using independent measures of staff and resident participation in treatment decisions, Holland and his coworkers (1981) found that job satisfaction among staff was not related to either measure, but that psychiatric patients improved more in programs in which patients and staff participated in treatment decisions and in which staff expressed more job satisfaction. Agbayewa, Ong, and Wilden (1990) noted that, despite initial resistance, staff were more satisfied when long-term care residents participated more in unit management. In general, better resident functioning and policies that permit resident choice and control promote staff and patient involvement and thereby tend to enhance work attitudes among staff. Such attitudes then have positive consequences for job satisfaction among staff and for functioning and behavior among residents (Moos & Schaefer, 1987).

Psychiatric Facilities

Our person-environment congruence framework (chapter 1) predicts that individuals with intact functioning respond positively to independence and freedom but that functionally and cognitively impaired people respond more to the degree of structure and support provided by the environment. A high degree of predictability and clear standards of behavior may be particularly important to those with cognitive impairment.

Smothers (1987) focused on this issue by using four POLIF subscales and the Ward Atmosphere Scale (WAS; Moos, 1989) to examine the relationship between program policies and behavior among relatively impaired long-term psychiatric patients. The patients showed more responsible behavior when they were in directive and structured programs; that is, programs that were high on expectations for functioning, did not tolerate problem behavior, and were low on policy choice. Low staff acceptance of problem behavior was also associated with more socially adaptive behavior.

Initial levels of patient impairment and chronicity were associated with the development of specific types of program policies. Specifically, staff tended to expect less functional competence and allow more problem behavior from chronic patients and from those who displayed dependence and maladaptive conduct (Smothers, 1987).

Overall, policies and services are especially important characteristics that reflect the underlying tone of a facility. Our findings show that these facility characteristics match residents' needs reasonably well, although the impaired residents in nursing homes give up some privacy and control for access to more health and daily living services. We return to the issue of the possible impact of different policy and service environments in chapters 8 and 9. In the next chapter, we focus on the social climates of residential facilities.

5

Social Climate

The social climate perspective assumes that every environment has a unique "personality" that gives it unity and coherence. Like people, some social environments are friendlier than others. Just as people differ in how they regulate their own behavior, settings differ in how they regulate the behavior of the people in them. Individuals form global ideas about an environment from their appraisal of specific aspects of it. These perceptions are based in part on reality factors; a judgment of friendliness might stem from whether residents greet each other in the lounge, help each other, participate in activities together, and so on.

The Sheltered Care Environment Scale (SCES), unlike the other parts of the MEAP, taps participants' subjective appraisal of their environment rather than objective information about the setting. The content of the SCES overlaps with the other measures but is more evaluative. For example, physical comfort is tapped by questions such as, "Is the furniture here comfortable and homey?" and "Is it ever cold and drafty here?" Resident influence is gauged by such questions as, "Can residents change things here if they really try?" and "Do residents have any say in making the rules?" Because the SCES measures the appraised environment, opinions are solicited from as many residents and staff members as possible; individual reports are then aggregated into a summary measure.

Integrated program assessment should cover such aspects of a setting as the quality of interpersonal relationships and the direction and strength of environmental press as the residents and staff evaluate them. For example, when nursing home residents were asked to identify areas they thought were especially important for good quality care, they emphasized

This chapter is coauthored by Bernice S. Moos.

the need for positive staff attitudes toward residents, policies that provide independence and personal control, a diversity of social and recreational activities, and a pleasant, homelike atmosphere (Holder, Frank, & Spalding, 1985).

Settings with similar objectively measured characteristics may nonetheless develop distinctive social environments and consequently have quite different impacts on residents and staff. For example, settings with policies that enable residents to participate in facility governance (high Resident Control as measured by the POLIF) should be seen as high on resident influence (as measured by the SCES); however, a facility with such a governance structure may be appraised as low on resident influence if residents' opinions and preferences actually have little or no impact on staff decisions and facility policies. Thus, although the social climate stems in part from other facility characteristics, such as physical features and policies, it also includes aspects of the facility that are not covered by other parts of the MEAP. In addition, the social climate can mediate or moderate the impact of objective environmental features.

Only a few investigators have tried to measure the social climate of residential facilities for older people. In order to test her person–environment congruence model, Kahana (1982) developed a questionnaire to assess environmental opportunities for particular behaviors as reported by staff members. The seven dimensions measure environmental stimulation, emotional expression, impulse control, level of autonomy and privacy, institutional control, organizational structure, and homogeneity or sameness. In one study, four of the dimensions differentiated among three homes for the aged (Kahana, Liang, & Felton, 1980).

Nehrke and his colleagues (1981) developed the Environmental Perceptions, Preferences, and Importance Scale (EPPIS) to assess 15 dimensions of the perceived environment. The EPPIS taps such areas as resident social interaction and stimulation, staff respect for residents, resident autonomy and choice, support for religiosity, responsive health care, physical barriers to mobility, and change versus sameness. Some of these dimensions were predictably related to morale, life satisfaction, and self-esteem among residents in a veterans domiciliary and a community residential care facility.

Although the social climate approach has not been applied widely in residential settings for older people, it has been used productively in a variety of other settings, including psychiatric treatment programs, health care and other work environments, residential settings for college students, and families. Three sets of social climate dimensions have been identified. *Relationship* dimensions assess the quality of interpersonal relations. *Personal growth* or *goal orientation* dimensions measure the directions in which personal development tends to occur in the setting. *System maintenance and change* dimensions deal with the degree to which the environment is orderly, clear in expectations, and responsive to change (Moos, 1987).

Working within this framework, we developed the SCES to provide administrators and staff of congregate residences with a relatively easy way to assess a facility's social climate. We wanted to formulate a method for assessing group living facilities by asking those who live or work there about the usual patterns of behavior. To be useful, such a measure must discriminate between facilities and reflect shared experiences that are subject to common agreement.

In this chapter, we describe the seven dimensions of the SCES and summarize normative SCES data from our sample of community facilities. (Information on the veterans facilities is presented in chapter 6.) We also examine the relationship between individual characteristics and SCES responses in order to address important issues in the interpretation of SCES results; these issues include the significance of aggregate scores and the effects of sample bias on SCES findings. Finally, we show how the SCES can be used to describe residential programs and assess how well they are implemented.

ASSESSING SOCIAL CLIMATE

The Sheltered Care Environment Scale (SCES)

The SCES is composed of 63 yes/no items that assess seven dimensions of the social climate of residential settings for older people. In general, the information is obtained from residents and staff. Each subscale comprises nine items (Table 5.1).

The first two subscales assess relationship dimensions. Cohesion measures how involved residents are with each other and how supportive staff members are toward residents. Conflict taps the extent to which residents express anger and are critical of each other and of the facility.

The second set of SCES dimensions assesses personal growth or goal orientation. Independence taps how self-sufficient residents are encouraged to be in their personal affairs and how much responsibility and self-direction they exercise. Self-Disclosure reflects the extent to which they openly discuss their feelings and concerns.

The third set of SCES dimensions measures system maintenance and change. Organization measures how important order and regularity are in the facility, whether residents know what to expect in their daily routine, and the clarity of rules and procedures. Resident Influence evaluates the degree to which residents can alter facility rules and policies and the extent to which staff restrict residents through regulations. Physical Comfort taps the level of comfort, pleasant decor, and sensory satisfaction provided by the physical environment. (For information on the normative and psychometric characteristics of the SCES, see Lemke & Moos, 1987; Moos & Lemke, 1992f.)

Table 5.1 Sheltered Care Environment Scale (SCES) subscale descriptions and item examples

RELATIONSHIP DIMENSIONS	
1. Cohesion	How helpful and supportive staff members are toward residents and how involved and supportive residents are with each other (Do residents get a lot of individual attention?)
2. Conflict	Extent to which residents express anger and are critical of each other and of the facility (Do residents ever start arguments?)
PERSONAL GROWTH DIMENSIONS	
3. Independence	How self-sufficient residents are encouraged to be in their personal affairs and how much responsibility and self-direction they exercise (Do residents sometimes take charge of activities?)
4. Self-Disclosure	Extent to which residents are encouraged to express openly their feelings and personal concerns (Are personal problems openly talked about?)
SYSTEM MAINTENANCE AND CHANGE DIMENSIONS	
5. Organization	How important order and organization are in the facility, extent to which residents know what to expect in their daily routine, and clarity of rules and procedures (Are activities for residents carefully planned?)
6. Resident Influence	Extent to which the residents can influence the rules and policies of the facility and are free from restrictive rules and regulations (Are suggestsions made by the residents acted upon?)
7. Physical Comfort	Extent to which comfort, privacy, pleasant decor, and sensory satisfaction are provided by the physical environment. (Can residents have privacy whenever they want?)

Each subscale comprises nine items.

Social Climate in Different Types of Facilities

Normative data are available for the three subsamples of facilities in which the SCES was developed (results for the veterans sample are presented in chapter 6). SCES data were obtained from more than 1,900 residents and 2,000 staff in the 135 nursing homes, from more than 1,200 residents and 390 staff in the 60 residential care facilities, and from more than 2,900 residents and 270 staff in the 67 congregate apartments. We averaged individual scores to produce two scores for each facility, one for residents and one for staff. Means of the subscale scores for each of the three sets of facilities are reported for residents in Figure 5.1 and for staff in Figure 5.2. These mean scores represent the average percent of items (each dimension has nine items) answered in the scored direction. Thus, on average, residents in nursing homes answered 62% of the items on the

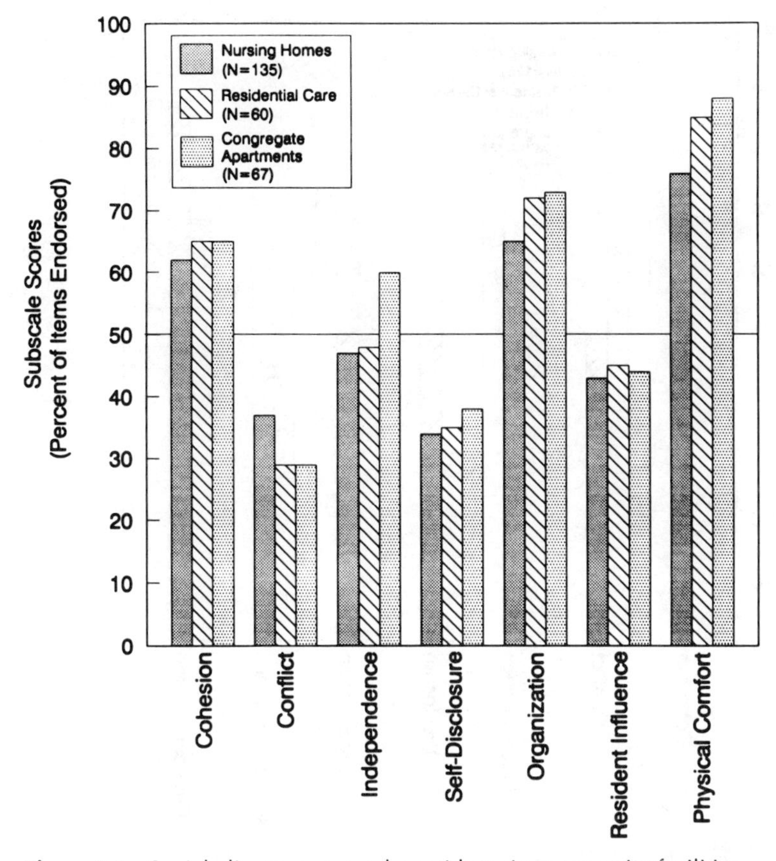

Figure 5.1 Social climates as seen by residents in community facilities.

Cohesion subscale to indicate that the particular aspects of cohesion were present in their facilities.

The general characteristics that differentiate facility types are reflected in their social climates. Compared with apartment residents, nursing home residents report higher conflict and lower independence, organization, and physical comfort. In the view of their residents, residential care facilities are similar to nursing homes in their lower levels of independence and similar to apartments in their lower level of conflict and higher levels of organization and physical comfort. Residents in all three facility types report similar levels of cohesion, self-disclosure, and resident influence. In general, staff responses show the same pattern of differences between facility types.

These findings show that the SCES discriminates between types of residential settings; five of the subscales for residents and six of the subscales for staff differentiate significantly between facility types. The SCES also discriminates among facilities that offer the same level of care (Lemke & Moos, 1987; Moos & Lemke, 1992f).

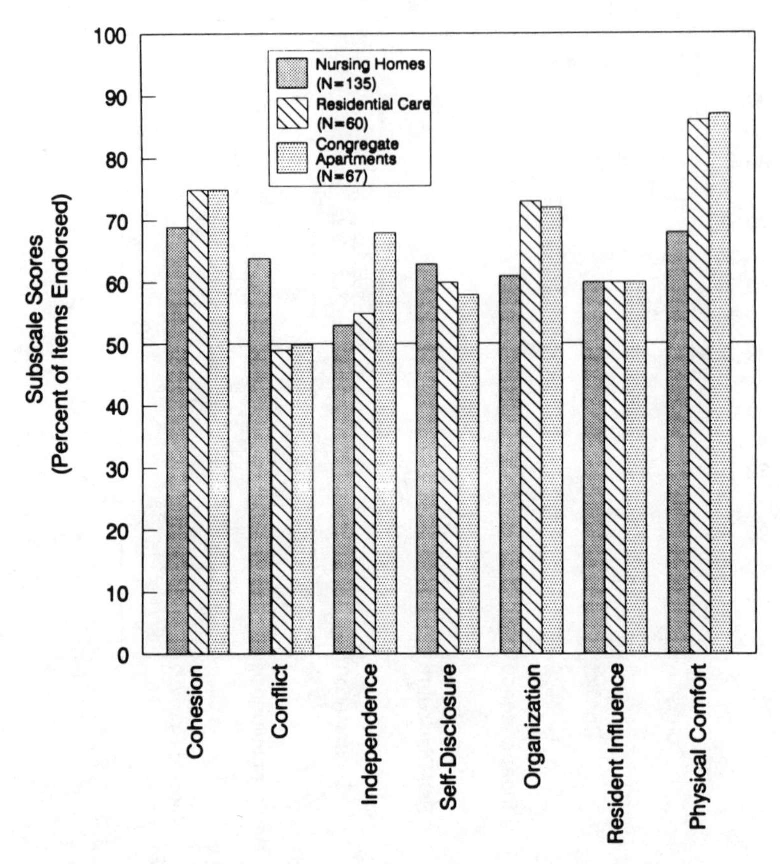

Figure 5.2 Social climates as seen by staff in community facilities.

PERSONAL FACTORS AND ENVIRONMENTAL PERCEPTIONS

Perceptions of social climate are the outcome of a process of concept formation, which stems from the interplay between actual events and qualities of a setting, the individual's role within the system, and the individual's personal values and beliefs. Although cognitive appraisal is generally based on reasonably accurate perceptions of environmental characteristics, individuals can have varied experiences in a given setting and may be predisposed to construe reality in terms that are compatible with their needs.

To the extent that environmental perceptions reflect personal characteristics of the respondent, social climate scales could be regarded as measures of personality traits tapped by items that happen to pertain to the social milieu; aggregating such scores from individuals would tell us little about the environment. Therefore, to interpret facility scores accurately, we need to know how much individual social climate perceptions reflect attributes of individuals rather than aspects of settings. Some evidence indicates that sociodemographic and other personal attributes are only minimally related to environmental perceptions (Moos, 1974, 1987;

Moos & Bromet, 1978). However, researchers have conducted little empirical work on this issue with older people.

To examine the contribution of individual characteristics and social role to the individual's perception of social climate, we obtained additional information from respondents in some facilities. A sample of 1,428 residents in 42 facilities completed the Background Information Form (BIF), which contains questions about residents' sociodemographic characteristics, current functional ability, and participation in organized and informal activities. (For more information on the sample and measures, see chapter 9 and Lemke & Moos, 1987.)

These data were used to examine the individual correlates of social climate perceptions. Partial correlations were computed between individual characteristics and the individual's SCES scores, controlling for the facility mean SCES score on the corresponding dimension. These analyses indicate the extent to which individual characteristics are systematically related to perceptions of a residential setting after the contribution of the actual environment, as measured by the consensual or average rating, is considered.

Residents' Personal Characteristics and Social Climate Perceptions

The results for residents indicate that an individual's perception of a residential facility is related to the average group perception of that setting and is relatively independent of personal characteristics. Correlations between individual perceptions and the facility average ranged from .34 for Self-Disclosure to .62 for Physical Comfort (average r = .50). Overall, variation in the average perception accounted for 26% of the variance among residents in their scores, whereas the individual characteristics added at most 2% of explained variance for individual SCES scores.

Some personal characteristics did have a small but significant relationship with social climate perceptions. For example, women tended to rate a given setting lower on physical comfort than did men. Better-educated residents reported that residents engage in more self-disclosure and have more influence in the setting. Residents who were more active reported somewhat more cohesion, independence, self-disclosure, and physical comfort, and residents who were better functioning tended to see the facility as more comfortable. Residents who lived in the setting longer tended to see it as less cohesive and less comfortable; residents seem to experience slightly greater isolation and dissatisfaction with longer tenure, a finding that is consistent with the pattern noted by Retsinas and Garrity (1985) for nursing home residents.

Staff Members' Personal Characteristics and Social Climate Perceptions

A sample of 985 staff members in 59 facilities provided information about their age, sex, and length of employment in the facility when they answered the SCES. The pattern for staff is similar to that for residents.

Individual staff perceptions of the facility social climate show a strong relationship to average staff perceptions and relative independence of individual characteristics. Correlations between individual scores and the facility average ranged from .42 to .53. Personal characteristics added at most 5% to the explained variance in individual SCES scores. Similarly, Orchowsky (1982) found little or no association between staff members' age or length of employment in nursing homes and their perceptions of the facility social climate.

As with residents, some personal characteristics were significantly related to social climate perceptions. Women staff members tended to report a higher level of self-disclosure in the environment than did men. Older staff members, who had a more positive perception of the setting than did younger staff members, reported somewhat higher cohesion, organization, and physical comfort and somewhat less conflict and self-disclosure. Staff who worked in a facility for a longer period of time reported more resident influence.

The Reality of Social Climate

Consistent with the results in other settings, we found some evidence that residents who were better functioning had somewhat more positive views of the setting. However, the overall findings show that the SCES reflects not just characteristics of the respondent but actual, agreed-upon qualities of a setting. For both residents and staff, the single best predictor of an individual's SCES scores is the group's perception. Knowing individual characteristics of the respondent does little to improve these predictions. Accordingly, the aggregate results reflect a core of consensual reality and indicate that findings should be relatively stable across subgroups of individuals.

The data on these issues are preliminary because we measured only a limited domain of individual characteristics. In particular, we did not include measures of resident or staff personality variables. However, studies in other types of settings indicate that personality traits show only weak relationships with social climate perceptions (Moos, 1987).

SOCIAL CLIMATE AND THE ADEQUACY OF PROGRAM IMPLEMENTATION

Profiles of residents' and staff members' perceptions of the social climate provide a picture of how these two groups see a facility. As with other parts of the MEAP, SCES standard score profiles (with a mean of 50 and standard deviation of 10) indicate how the facility ranks when compared with similar settings. We present three examples here from the facilities described in chapters 3 and 4. (For additional examples of SCES profiles, see Lemke & Moos, 1985; Moos, Gauvain, et al., 1979; Moos & Lemke, 1983, 1992f; Moos, Lemke, & Clayton, 1983.)

Middleton Nursing Home

The social climate of Middleton is shown in Figure 5.3, which compares resident responses with the norms for nursing home residents and staff responses with the norms for staff. Both residents and staff give fairly typical ratings of the cohesiveness in relationships among residents and between residents and staff; for example, nearly all residents and staff said that requests are taken care of right away and that staff do not talk down to residents. Both groups report above-average expression of disagreement and anger among residents (high conflict).

Although residents and staff agree on the quality of relationships, these two groups see the program emphasis somewhat differently. Residents see little evidence of independence and about-average openness in discussing personal concerns. In contrast, staff give higher ratings on both of these dimensions.

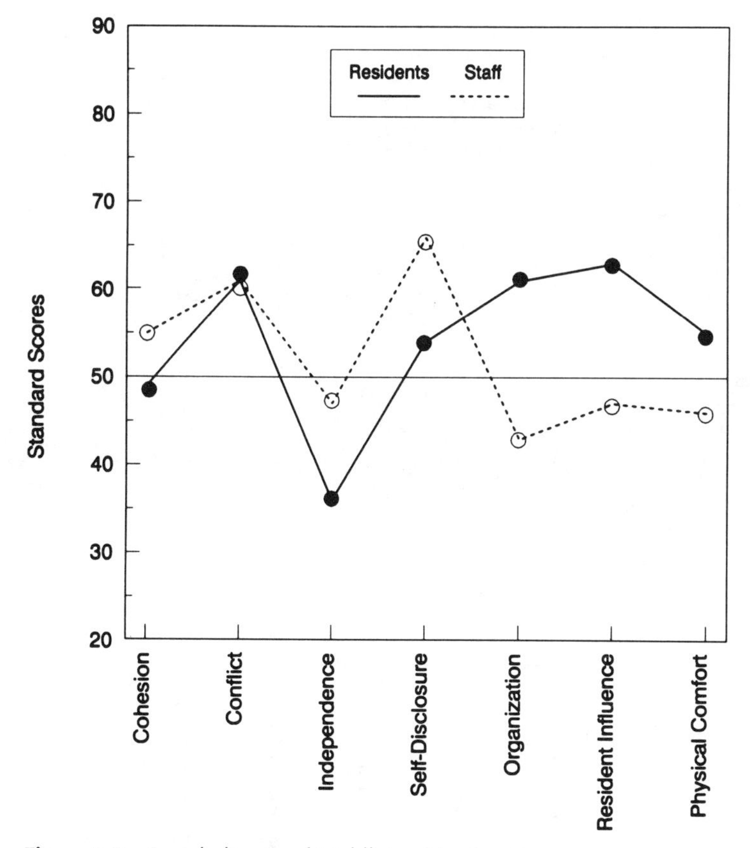

Figure 5.3 Social climate of Middleton Nursing Home.

Residents and staff also disagree in their descriptions of the facility's system maintenance and change dimensions. For example, residents rate organization much higher than do staff and higher than average. To residents, the facility seems well run and efficient. In contrast, staff report that policies are unclear and always seem to be changing. Residents also see themselves as having a significant voice in facility affairs, whereas staff rate residents' influence as about average. Residents were more likely than staff to say that a resident would not be asked to leave for violating rules or fighting with another resident; the staff view is more consistent with the policies reported on the POLIF, which reflect a lower than average tolerance for problem behavior (see Figure 4.2).

Maude's Community Care Home

Residents and staff of this small community care home disagree in how they see the social climate (Figure 5.4). Residents see the level of support as about average, but staff see relationships as quite cohesive. For example, the staff all said that they spend a lot of time with residents, that

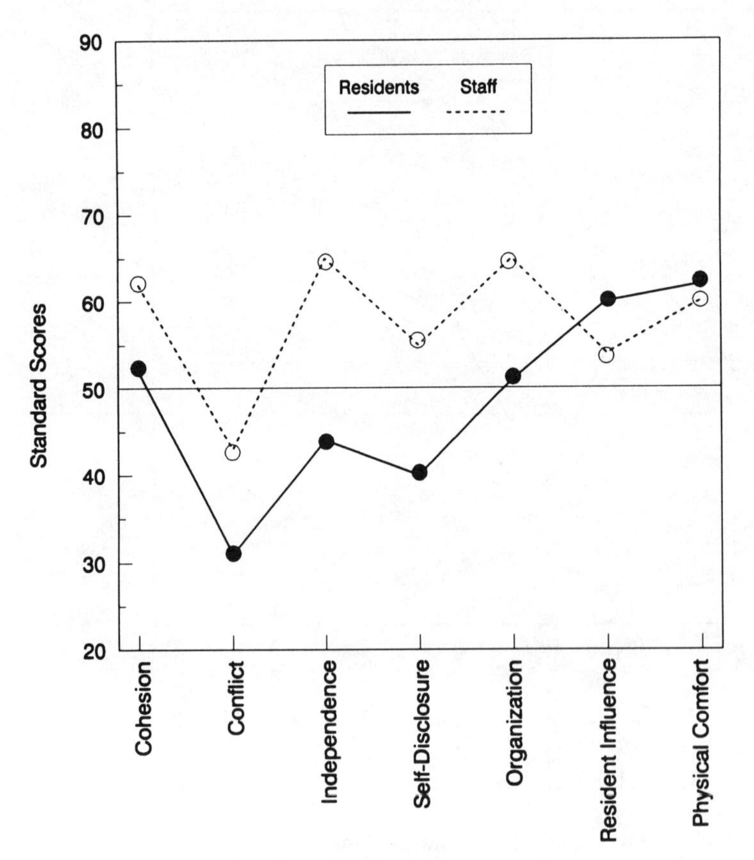

Figure 5.4 Social climate of Maude's Community Care Home.

there are many social activities, and that requests are taken care of right away, whereas only a bare majority of residents endorsed these items. Staff members report a moderate level of conflict; a majority of staff said that residents sometimes start arguments, argue openly, and express their anger. However, residents were nearly unanimous in reporting a very low level of conflict.

Residents see about-average emphasis on independence and slightly below average emphasis on self-disclosure, whereas staff see a stronger emphasis on both these personal growth dimensions.

The two groups also differ in their evaluation of the facility's organization. Staff report that it is efficiently organized and well run, but residents rate this dimension average. In spite of policies that permit only low levels of choice and control (Figure 4.3), residents feel that they can influence how the facility is run; in the absence of formal policies, as indicated by the score on POLIF Resident Control, the facility's small size may allow residents to feel that they can exercise influence through informal channels. In contrast, staff members, who rate resident influence as about average, are somewhat skeptical about residents' ability to cause change.

Amity House

Residents and staff show marked agreement in their views of the social climate of Amity House (Figure 5.5). Both groups report that relationships are supportive and free of conflict and that residents have a high level of independence, but a relatively low level of self-disclosure.

Organization is reported to be about average, but resident influence is very high, consistent with the stated policies giving residents a high level of control (see Figure 4.4). Both groups of respondents agreed that the staff members are not strict about rules and regulations, that new and different ideas are often tried out, and that staff act on residents' suggestions.

Program Climate and Standards of Implementation

Just as with the PAF and POLIF, the SCES can be used to assess how well a program has been implemented and to guide change efforts.

Middleton Nursing Home

The findings for Middleton suggest two areas that could be the focus of interventions: the high level of conflict and the low level of independence experienced by residents. One challenge is to maintain openness and free expression, as indicated by the moderate to high score on self-disclosure, while reducing the strain that accompanies high conflict. This challenge might be met by allowing residents to express their personal concerns and criticisms in a supportive context, such as in small-group discussions conducted by an experienced leader. Intergenerational activities can provide another opportunity for free expression in a context of closeness and pro-

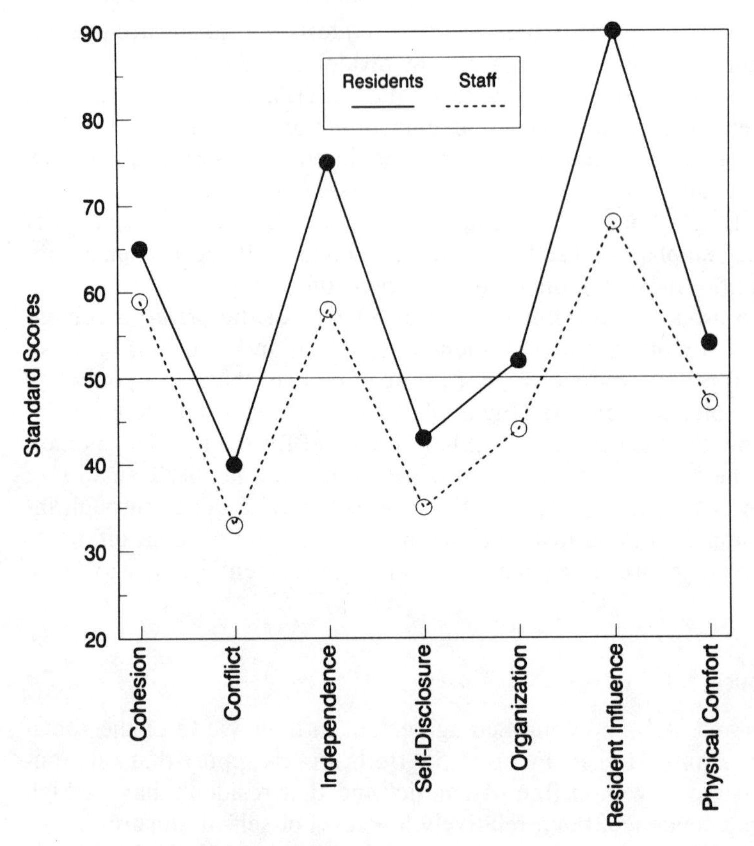

Figure 5.5 Social climate of Amity House.

ductiveness. A secondary benefit of such activities could be increased cohesion.

The policies of Middleton's administration emphasize a high level of choice for residents but offer few formal avenues for resident control (see Figure 4.2). Residents report that they can influence how the facility is run, but they see little emphasis on independence. Here the challenge is to maintain the facility's informal responsiveness to residents and to structure opportunities for residents to make contributions in spite of their impairment. In-service training may help sensitize staff to the subtle differences between forms of assistance that undermine residents' sense of independence and those that support it.

Maude's Community Care Home

Residents and staff in Maude's Community Care Home hold quite different views of the social climate; these discrepant perceptions provide a focus for intervention efforts. Staff need to be more aware of how residents experience the setting. Such awareness may help them gauge the potential impacts of their actions and balance their somewhat idealized image of the facility climate. Residents report moderate levels of interpersonal engagement; they do not argue, but neither do they feel particularly close to each other or free to discuss personal concerns. Providing mutual support or discussion groups might help to increase the level of cohesion and self-disclosure in the facility.

As is the case for Middleton, Maude's residents feel that they can influence how the facility is run but have only moderate independence. In contrast with Middleton, the stated policies allow residents few opportunities for choice even though residents are able to carry out most activities without assistance. Thus, a first step to increase the level of independence may be to give residents more latitude in determining their own daily routines.

Amity House

In contrast with Maude's and Middleton, Amity House is highly cohesive and oriented toward fostering residents' independence and influence, a view shared by residents and staff. This social climate profile suggests that Amity House is an effectively implemented program of congregate housing and could serve as a useful model for other programs.

Evaluating Other Facilities

Several researchers have used the SCES in a similar manner to assess program implementation. Waters (1980a) found that residents and social services staff in a nursing home reported high cohesion and independence but that nursing staff reported only average emphasis in these and other areas. When asked about their preferences, staff members wanted less conflict and more emphasis on cohesion, independence, and organization. Waters (1980b) described uses of such information to plan and evaluate program change (for more details, see Moos & Lemke, 1992a). Other researchers have used the SCES to assess program implementation in congregate apartments and long-term care facilities (Billingsley & Batterson, 1986; Ramian, 1987; Shadish et al., 1985; Thompson & Swisher, 1983).

Evaluating Program Models

In comparing programs against normative standards, researchers and practitioners should keep in mind the guiding conceptual model for the program. Thus, for example, a community care program can be considered well implemented from a conceptual perspective if the program is oriented toward rehabilitation or treatment. In contrast, Blake (1987) found that a sample of group homes in New Jersey was characterized by a mechanistic and controlling social climate and concluded that the model was poorly implemented.

Cross-Cultural Standards of Program Implementation

Both normative and conceptual implementation standards may vary in different cultures. For example, Benjamin and Spector (1990) found that a small group of British facilities scored somewhat lower than American facilities on independence, self-disclosure, organization, and physical comfort. Similarly, when Netten (1991; 1991–92) asked staff members to assess the social climate in British residential homes, she found that the British homes were lower than American nursing homes on cohesion, independence, and organization and higher on conflict and resident influence.

Fernandez-Ballesteros and her colleagues (1991) found that residents and staff in Spanish residential care facilities saw much more conflict and much less cohesion and physical comfort than did residents and staff in American facilities. Residents in the Spanish facilities also reported more self-disclosure, and staff reported lower organization. According to the authors, these findings may reflect differences between the Spanish and American social context; for example, Spaniards may be more expressive and therefore engage in more self-disclosure and conflict, as well as in more open criticism of the facilities' physical comfort. More important, these findings illustrate how normative standards of program implementation may vary in different countries.

A Typology of Social Climates

The information obtained with the SCES, when organized so as to identify general types of residential programs, can help identify different routes to effective program implementation. Such a typology may assist in placing residents in programs that match their needs. With these purposes in mind, we identified six distinct types of social climates in residential facilities, including two types providing somewhat different models of well-implemented programs (Timko & Moos, 1991b). One type, comprising a cluster of supportive, self-directed facilities, is characterized by high scores on cohesion, independence, and resident influence and low scores on conflict; Amity House is an example of such a facility. Facilities in a second cluster also emphasize supportive interpersonal relationships and organization but are less oriented toward independence and resident influence. The other four types are characterized by open conflict, suppressed conflict, emergent-positive, and unresponsive social climates.

RESEARCH AND POLICY APPLICATIONS

A number of research issues can be addressed when the SCES is used to measure a facility's social climate. Chapters 6 and 7 focus on how size, ownership, and other facility characteristics influence the social climate of a living group. Here we consider research on resident–staff agreement, the relationship between social climate and observers' evaluations of facility quality, the social climate of hospital and community care, and the link between social climate and residents' morale and well-being.

Resident–Staff Agreement

Residents' perceptions provide an essential perspective on residential facilities (Golant, 1986; Lehman, 1983; Wilkin & Hughes, 1987). Information about living groups can be obtained from outside observers, who may be more objective, but residents have time to form accurate, durable impressions of a facility's social climate because they live in the facility and experience it in an ongoing way. Although staff have a somewhat different perspective because of their role in the facility, they too have ample opportunity to form durable impressions of the social climate.

We found that residents and staff agree moderately well about the social climate of their facility; the average correlation of the SCES means between residents and staff across the community facilities is .41. However, the extent of resident–staff agreement varies from one facility to another (see Figures 5.4 and 5.5). Accordingly, we recommend that researchers and practitioners obtain the views of both groups and consider the extent of agreement between the groups as an additional source of information about the facility.

Resident and staff groups may develop different perceptions because of the two groups' distinctive roles in the facility and differences in their value systems and frames of reference. For example, people who have more authority in an environment, such as staff members, tend to see it more positively than those who have less authority (Moos, 1989). In this vein, staff in all three levels of care tend to report more cohesion, independence, and resident influence than do residents (Figures 5.1 and 5.2). Higher-status staff, such as nurses, also report more resident influence than do lower-status staff. However, staff also report more conflict than do residents, and staff in nursing homes are more critical than residents of the facility's organization and physical comfort. Thus, staff do not have a simple positive-response bias. Other researchers have reported similar findings when comparing the views of staff and residents (Smith, 1984; Stein, Linn, & Stein, 1987).

Items from the Cohesion subscale can be used to illustrate the ways in which staff and residents' roles in the facility influence their perceptions. Staff members are more likely than are residents to say that residents get individual attention and that staff spend time with residents. From the standpoint of individual staff members, nearly all of their time may be given to meeting residents' needs. However, interaction with staff members may constitute only a small fraction of the total time a resident spends in a facility. Thus, differences in the two groups' perceptions of cohesion reflect in part real differences in their experience of the setting. Staff members are also likely to be exposed to a broader range of interactions than are residents, and, consequently, are more likely to report instances of conflict and self-disclosure.

In addition to staff and residents, other observers can provide a valuable perspective on the setting. For example, Jones and Batterson (1982) gave the SCES to family members of nursing home residents. Comparison of family, resident, and staff perceptions indicated that, on average, family members agreed more closely with residents than with staff. The views of family members may be particularly useful in facilities where residents are too impaired to complete the SCES themselves.

Observers' Ratings of Quality

Comparisons of social climate perceptions and other assessments of facility quality help establish the validity of the SCES and contribute to our understanding of the components of high-quality care. Stein and her colleagues (1987) obtained SCES ratings from residents and nursing staff in 10 community nursing homes. Social workers with special training in evaluating nursing homes also independently rated the quality of the 10 homes. In the homes judged to be of better quality, residents and staff reported more cohesion and independence. Residents in the better homes also reported better organization, more resident influence, and greater physical comfort than did residents in the poorer homes.

We had outside observers rate the pleasantness of each facility in the community sample in terms of how willing they would be to place a person there. Ratings of pleasantness were associated with higher levels of cohesion, organization, resident influence, and physical comfort as reported by both residents and staff. In addition, facilities rated by observers as more attractive and as offering more diversity were seen as more comfortable by residents and staff (see chapter 3).

According to our findings and those of other researchers (e.g., Stein et al., 1987), residents' perceptions of social climate are more closely related to observers' ratings of pleasantness than are staff members' perceptions. These findings show that lucid residents can give valid feedback on social climate and that an assessment of social climate is one good way to evaluate the quality of nursing home care (see also Ramian, 1987). (For a discussion of some aspects of the construct validity of the SCES, see Lemke & Moos, 1990; Smith & Whitbourne, 1990).

Hospital and Community Care

Several researchers have shown that residential facilities tend to provide more harmonious and flexible living situations than those typically found in medically oriented facilities. Maloney and Bowman (1982) used the SCES to compare the living situations of psychiatric patients released to group homes with the environment of patients who remained in a hospital-based treatment program. In general, residents and staff agreed that the group homes were higher on cohesion, independence, organization, and physical comfort and lower on conflict than the hospital program. Residents and staff in the group homes showed closer agreement on the social climate than did residents and staff on the hospital unit.

Other researchers have obtained similar findings. Lehman, Possidente, and Hawker (1986) examined the quality of life experienced by chronically mentally ill persons who lived either in a psychiatric hospital or in supervised community programs. Compared with hospitalized patients, community residents described their living situations as more cohesive and comfortable and more oriented toward independence and resident influence. Community residents were also more satisfied with their living situation; in turn, satisfaction with the living situation was associated with overall life satisfaction. The findings show that the quality of mentally ill patients' housing is an important component of their quality of life (Lehman, 1988).

Although community-based programs for chronically mentally ill people often provide a positive living environment, we noted earlier that the policies of some community care homes are not consistent with the principle of normalization (chapter 4; Blake, 1985–86, 1987). Some community facilities provide only a custodial setting with little or no treatment orientation (Moos, 1988b).

Social Climate and Residents' Morale and Well-Being

Process analyses focus on the causal chain between an intervention program and outcome. One link in that chain is the relationship between the program as actually implemented and clients' morale and well-being. As noted in our model (Figure 1.1), the social climate may play a unique role in determining the effectiveness of a program. Social climate mediates some of the influence of physical features and policies and has an independent effect on outcome. We focus on these processes at the program and individual level of analysis and then review some studies of person–environment congruence and resident morale. We examine program-level relationships in more detail in chapter 8.

Facility Social Climate and Residents' Adaptation

As noted earlier, we identified six distinct types of facility social climates (Timko & Moos, 1991b). Residents in facilities with supportive self-directed and supportive well-organized social climates were rated as higher on resident adjustment, participated in more self-initiated activity in the facility and in the community, and relied less on the health services offered in the facility than did residents in the other types of social climates. Residents in facilities that were high on conflict and low on each of the other social climate dimensions tended to do worse on each of these outcome criteria.

Netten (1991, 1991–92) conducted a cluster analysis of staff responses to the SCES in British residential homes and identified a set of homes that were cohesive and well-organized and oriented toward independence, self-disclosure, and resident influence. These homes were associated with better resident outcomes. In an earlier study, Netten (1989) also found some positive concomitants of conflict: Residents in homes that were higher on conflict were better oriented to their surroundings, perhaps because conflict is associated with the facility layout or with residents' activity levels. More generally, the level of conflict should be considered in conjunction with other aspects of social climate, such as the levels of cohesion and self-disclosure.

Individual Residents' Perceptions and Well-Being

Using an individual level of analysis, several researchers have examined the associations between residents' perceptions of nursing home environments and their mood and well-being. Shadish and his colleagues (1985) found that residents who viewed their nursing home as cohesive and low in conflict, as oriented toward independence and resident influence, and as well organized and comfortable tended to report higher well-being. Clark (1989) obtained basically similar findings. Shadish and his coworkers (1985) also noted that residents' well-being generally was not related to their levels of symptomatology and social integration; the strongest connections were to their appraisal of the quality of the nursing home environment.

Colling (1985) examined the associations between nursing home residents' perceptions of the social climate, their perceived control over daily living activities, and their well-being. Residents who reported more cohesion in their nursing home experienced more control over activities of daily living and, in turn, higher well-being. Residents who develop more supportive relationships with staff may feel that they have more control over their daily activities, and such personal control is likely to contribute to their well-being. Colling also found that residents who appraised their facility as more comfortable reported higher well-being.

In a study of community care homes, Namazi and his colleagues (1989) noted that, aside from self-perceived health, the best predictors of residents' psychological well-being were residents' perceptions of comfort in the home, comfort with other residents, and the quality of interaction among residents. A familylike atmosphere in the home and personalization of care also predicted well-being.

Because these studies focused on individual residents, the relationships identified may reflect processes operating within the individual. For example, the findings may reflect a tendency for optimistic or resilient individuals, with the potential for better adjustment, to see their facility in a more positive light. More specifically, residents with a more positive outlook may elicit more support from other residents and staff and participate more actively in facility programs; as a consequence, these optimistic, active residents may develop more positive views of the social climate and experience more favorable outcomes. In earlier longitudinal research, we found that patients who initially held more positive views of their psychiatric program have better short- and long-term treatment outcome at follow-up (Finney & Moos, 1984, 1992). Thus, whatever their source, program perceptions may provide important predictive information about individuals' later morale and well-being.

Person–Environment Congruence and Resident Morale

In addition to specific aspects of social climate, which directly contribute to improved resident outcomes, the congruence between social climate and individual preferences and needs also can influence morale and well-being. Kahana and her colleagues (1980) measured nursing home residents' preferences and the facility social climate (as assessed by staff) on seven commensurate dimensions (see above) and found both direct and congruence effects. Residents in facilities that provided more environmental stimulation were better adjusted than residents in facilities that were low in stimulation. Thus, environmental stimulation enhanced residents' morale irrespective of their personal preference for stimulation.

In several other areas, however, person–environment congruence explained a significant increment in residents' morale after personal and social climate variables were considered. For example, residents whose level of impulse control fit with the level of control provided by their environment reported higher morale than did residents who preferred either more or less impulse control than their environment provided. Thus, for impulse control, a deviation from residents' preferences in either direction was associated with lower morale.

A more common situation is that residents who want more environmental resources than their environment provides tend to experience lower morale. Thus, Kahana and her colleagues (1980) found that residents who preferred more autonomy and privacy than were available in their facility had lower morale than those who wanted less autonomy and privacy than were provided. Similarly, residents who preferred more variety and individualized treatment than they had in their facility reported lower morale than those who wanted more homogeneity than the facility offered.

Similarly, Orchowsky (1982) examined person–environment congruence in determining resident outcomes. He gave the SCES to residents in three homes for older people and also asked them to rate the level of emphasis they wanted on each SCES dimension. The connection between residents' perceptions of social climate and their life satisfaction depended on their preferences. For example, perceptions of organization were more closely related to life satisfaction for residents who preferred a well-organized setting. Orchowsky (1982) obtained comparable results for conflict, self-disclosure, and physical comfort.

Nehrke and his colleagues (Nehrke et al., 1984; Sperbeck, Whitbourne, & Nehrke, 1981) relied on the agreement between an individual's perceptions and preferences to index person–environment congruence. In programs with more resident–staff interaction, residents generally appraised the environment as more congruent with their preferences; such congruence was positively related to residents' morale and life satisfaction. In a more restrictive environment, where residents had less personal autonomy or fewer behavioral alternatives, congruence was even more closely related to residents' well-being. In less restrictive environments, congruence between the appraised and preferred social climates may be less important because less restrictive settings give greater opportunity for residents with varied preferences to find an appropriate niche.

Coulton, Holland, and Fitch (1984) followed patients who were discharged from psychiatric hospitals into community care homes. Patients rated their personal characteristics and those of their home environment on a set of commensurate dimensions. Consistent with Kahana and her colleagues (1980), when patients' needs for privacy and self-understanding were not met, their functioning tended to decline. When environmental demands were too strong, that is, when the emphasis on task performance and autonomy exceeded patients' capabilities, patients showed less stability in functioning over time.

This research offers examples of situations in which congruence between the individual and the environment contributes to understanding and predicting residents' outcomes. Thus, residents who prefer more autonomy and self-disclosure appear to do better in facilities that emphasize these areas. Strong emphasis on performance and too much stimulation can lead to lower morale and well-being among some impaired residents, who are able to function better in a more tolerant and supportive setting. We return to these issues in chapters 8 and 9.

6

Ownership, Size, and Facility Quality

In preceding chapters, we focused on the physical features, policies and services, and social climate of community residential facilities for older adults. We showed that these characteristics vary substantially in different facilities, depending in part on their level of care. In this chapter, we examine two other determinants of facility quality: ownership and size. These two structural factors, which are associated with facility characteristics that define the quality of care, have important implications for residents. Here, we compare proprietary and nonprofit facilities with each other and with a sample of publicly owned facilities for veterans; we also consider the relationship between facility size and a facility's resources and social environment.

OWNERSHIP AND FACILITY QUALITY

The responsibility for providing residential services to the older population has shifted among different sectors of society in response to changes in the numbers and needs of older people and the economic resources available for their care. At various times, the public, private philanthropic, and proprietary sectors have expanded to fill these needs. Concurrently, researchers and practitioners have debated the impact of facility ownership on the quality of housing and services for older people.

The Profit Motive, Efficiency, and the Quality of Care

The general outlines of this dialogue have remained fairly constant. Critics of proprietary ownership have argued that in the inevitable conflict between making expenditures for better quality care and making a profit, quality of care will suffer; particularly where demand outstrips supply, as is the case for services to older people, the profit motive is thought to lead to deterioration in care or escalation of costs. Historically, this argument has served as a rationale for governmental regulation of such settings.

Nonprofit ownership has generally escaped such criticism. Proponents argue that nonprofit ownership combines the advantages of other forms of ownership without their disadvantages. Because nonprofit facilities need not allocate money to pay shareholders, a given health care dollar goes farther than in proprietary facilities. The organizational ethos is one of service to the consumer, a stance that may create fewer conflicts for the professional staff (Majone, 1984). Residence in a nonprofit facility often occurs within the context of long-term affiliation with a religious or fraternal organization; thus, nonprofit facilities may provide more sense of community and accommodation to individual needs.

Nonetheless, nonprofit facilities are subject to their own set of incentives, some of which may detract from good quality care. For example, managers in nonprofit facilities may not feel strong pressure to minimize costs and reduce wasteful administrative procedures, may be reluctant to admit residents who are more impaired or residents on Medicaid, and may lack the initiative to be innovative and seek out new markets (Weisbrod, 1989). Overall, nonprofit facilities may be less efficient than proprietary facilities (Caswell & Cleverley, 1983; Meiners, 1982; Ullmann, 1987). In addition, despite their supposed advantages, nonprofit organizations have failed to meet the growing demand for housing and services for older people.

With regard to public institutions, critics have argued that, despite humanitarian motives, government agencies provide poor quality care because they are not subject to the rigors of a competitive marketplace. According to this view, individuals within these systems are free to be inefficient and self-serving; two potential negative consequences are entrenched bureaucratic administration and lack of innovation.

Government agencies have been involved in direct provision of care to two groups of older people: the indigent and veterans. Some of the first facilities in this country to house exclusively older residents were those for veterans, whose service to the nation was viewed as entitling them to care in their old age. Recent debate over whether the Department of Veterans Affairs should provide these services directly or purchase them from the private sector reflects continuing interest in the impact of public ownership on the quality of care.

Prior Research on Ownership Types

Researchers investigating the impact of ownership on the quality of care have had mixed findings (Davis, 1991; Hawes & Phillips, 1986). In a large sample of British institutions, Townsend (1962) found that nonprofit homes had the most desirable attributes, proprietary nursing homes were intermediate, and public homes had the least desirable attributes. In contrast, Holmberg and Anderson (1968) found that proprietary and nonproprietary homes in Minnesota were quite similar; however, the investigators grouped together government and voluntary nonprofit homes, possibly canceling some differences. In another large survey of American nursing homes, Levey and his colleagues (1973) found no relationship between ownership (corporate proprietary, individual proprietary, and nonprofit) and such resources as the level of nursing staff, the adequacy of records, and the provision of physical features for safety, recreation, and rehabilitation.

Proprietary facilities may skimp on personnel costs in order to secure profits (Rango, 1982). Fottler, Smith, and James (1981) found that lower staffing levels were associated with higher profits in proprietary nursing homes. Comparing ownership types, Elwell (1984) found that nonprofit and government facilities allocated more financial resources to personnel costs and provided more staff hours. Greene and Monahan (1981) found that for-profit ownership was related to fewer professional nursing hours and lower expenditures on patient care, dietary services, and professional nursing care. Similarly, Ullmann (1987) found that expenditures for patient care were lower in for-profit than in nonprofit or government facilities; nonprofit facilities were also rated as somewhat higher on the quality of rehabilitation services.

Conceptually similar findings apply to psychiatric facilities and hospitals: Compared with for-profit hospitals, Arrington and Haddock (1990) found that nonprofit hospitals offer more social benefits in the areas of access to care, services provided, and staff training. Compared with nonprofit providers, for-profit mental health service providers devote fewer staff resources to patient care (Schlesinger & Dorwart, 1984) and provide fewer services (Culhane & Hadley, 1992). More generally, as Shadish (1989) noted, the shift from public- toward private-sector facilities has not improved the quality of care for chronic mental patients, in part because marketplace incentives motivate managers to house patients as cheaply as possible and because potential protective mechanisms, such as informed consumer choice and government regulation, are ineffective countervailing forces (but see Nyman & Geyer, 1989, for a different perspective on consumer choice).

Some researchers have questioned whether higher costs and increased expenditures translate into better quality of care (e.g., Nyman, 1988). For example, Schlenker and Shaughnessy (1984) analyzed data on Colorado nursing homes and found that their quality measures failed to account for the higher costs of nonprofit facilities. However, few studies have incorporated quality measures that do not directly reflect expenditures.

Although theoretical discussions and empirical investigations of the impact of ownership have tended to focus on health care settings, including nursing homes, the general arguments for the advantages of nonprofit facilities and for the disadvantages of proprietary and public ownership can be applied across levels of care. However, different levels of care vary in funding sources and regulatory environments, which may influence the impact of ownership.

We focus here on the differences among 147 proprietary, 91 nonprofit, and 81 veterans facilities. To examine the influence of ownership, we conducted analyses of covariance, controlling for level of care. The basic differences that we discuss here remain after variations in facility size are taken into account.

The analyses exclude 24 facilities (8 nursing homes, 3 residential care facilities, and 13 congregate apartments) in the community sample that are not under proprietary or nonprofit ownership and consist largely of county- or state-run facilities. Because of the small number and restricted diversity of these nonveteran public facilities, we rely on the veterans sample as the example of public ownership.

PROPRIETARY VERSUS NONPROFIT OWNERSHIP

In our sample, nonprofit facilities are about 50% larger than proprietary facilities (see Table 6.1). The two ownership types are similar in their location; about half are located in business areas or in neighborhoods with mixed business and residential uses. On average, these facilities were built in the early 1960s.

Table 6.1 Characteristics of facilities and residents in proprietary, nonprofit, and veterans facilities

	Proprietary (N = 147)	Nonprofit (N = 91)	Veterans (N = 81)
FACILITY CHARACTERISTICS			
Size (number of residents)	94	148	101
STAFF CHARACTERISTICS			
Number of staff per 100 residents	52	57	59
Women (%)	79	77	60
Employed more than 1 year (%)	56	64	78
Staff resources	45	56	64
RESIDENT CHARACTERISTICS			
Average age (years)	77	82	69
Women (%)	66	82	6
Marital status (%)			
Never married	16	14	27
Married	11	15	30
Divorced	11	4	24
Widowed	62	68	19
Nonwhite	9	9	10
In residence more than 1 year (%)	62	74	69

Analyses control for level of care. The three sets of facilities differ significantly ($p < .01$) on all variables except the proportion of nonwhite residents.

The overall staffing level is similar in the two ownership types, but nonprofit facilities have a higher ratio of professionally trained staff and a lower turnover rate. More of the staff members in nonprofit facilities have worked there for more than a year, perhaps in part because more of the staff are professionals. Although proprietary and nonprofit facilities have similar levels of staffing, the lower turnover and higher proportion of professional staff contribute to the higher score of nonprofit facilities on Staff Resources.

Resident Characteristics

The residents in nonprofit facilities are about 5 years older on average than residents of proprietary settings. Given their more advanced age, it is not surprising that they are more likely to be women. These individuals also are more advantaged than those in proprietary facilities; they are better educated and more likely to use personal savings to pay for their care.

Even though functional dependencies tend to increase with age, the average functional ability of residents in nonprofit settings is about the same as that of proprietary facility residents. Residents in the two groups of facilities also are comparable in their levels of involvement in self-initiated and community activities. Residents of nonprofit facilities appear to experience better health and functioning at a given age, perhaps due to their personal resources, which may also enable these individuals to remain longer in private housing.

Residents of nonprofit facilities have somewhat greater residential stability; more of them have lived in their present facility for longer than a year. Although the residents of nonprofit facilities are older, the mortality rate in these facilities is the same as that in proprietary facilities (averaging about 4% during the 3-month period preceding our assessment), as is the rate for other discharges (averaging about 9% over a 3-month period). Thus, compared with proprietary facilities, nonprofit facilities have a larger core of long-term residents who experience a comparatively low rate of turnover and a smaller group of new admissions who have a higher turnover rate. These findings are further evidence that, overall, residents of nonprofit facilities enjoy better health and functioning at a given age.

Physical Features

Proprietary and nonprofit facilities occupy buildings constructed at about the same time, but nonprofit settings provide their residents with a more comfortable physical environment (Figure 6.1). Nonprofit facilities have more physical features that add convenience and comfort (physical amenities and social–recreational aids), more features that provide a safe and secure environment (orientational aids, safety features), and more space for resident and staff functions (space availability and staff facilities). Consistent with these findings, outside observers using the Rating Scale

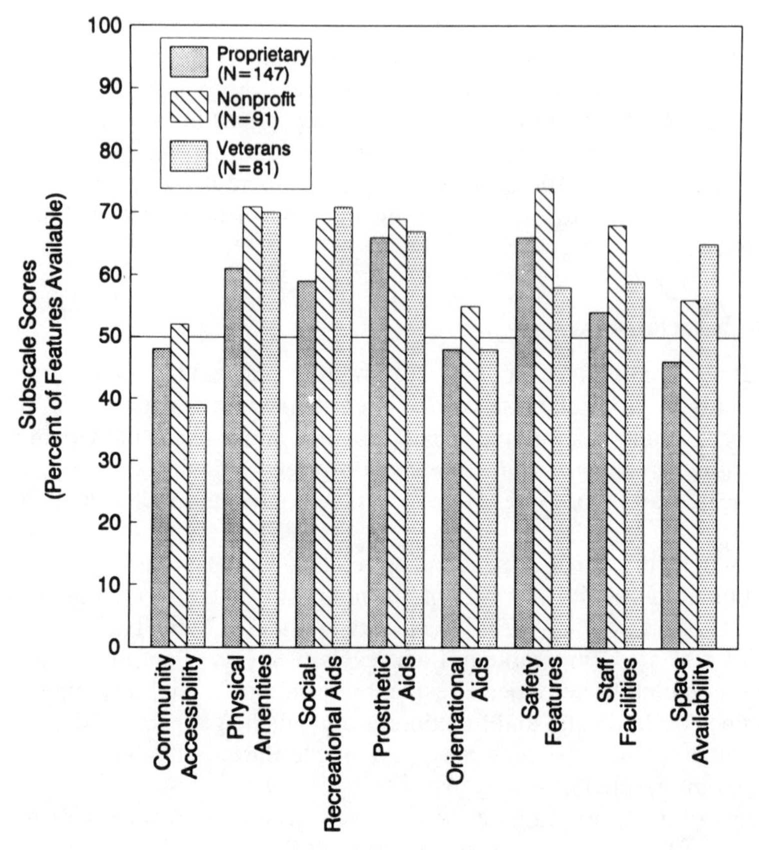

Figure 6.1 Physical features in proprietary, nonprofit, and veterans facilities.

described nonprofit facilities as more attractive and stimulating. Overall, compared with proprietary facilities, nonprofit facilities appear to invest more in their physical plant.

Policies and Services

Consistent with other investigators (Elwell, 1984; Greene & Monahan, 1981; Ullmann, 1987), we found that differences in services and staffing generally favor nonprofit facilities. For example, nonprofit facilities offer more health services (Figure 6.2), and as we noted earlier, they have a more stable, diverse, and well-trained staff. Even though nonprofit facilities provide more health services, their policies reflect slightly higher expectations of resident functioning. Nonetheless, nonprofit facilities are more likely to accept noncompliant behavior from residents.

Nonprofit facilities make more provisions for communicating with residents and staff and for giving residents a formal voice in running the facility. In addition, nonprofit facilities provide their residents with more privacy and somewhat more flexibility in scheduling. The findings suggest a difference in philosophy of care, with nonprofit facilities placing more

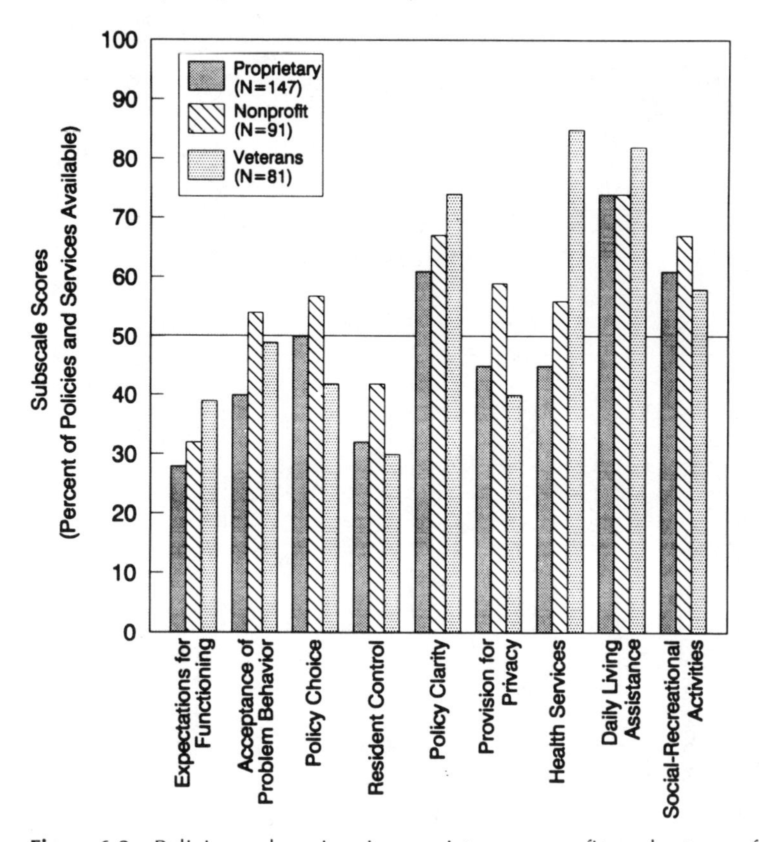

Figure 6.2 Policies and services in proprietary, nonprofit, and veterans facilities.

emphasis on recognition of the older person's need for a sense of independence and mastery. In contrast, the emphasis on efficiency in operating proprietary facilities may lead to more hierarchical decision making and less concern with individual control.

Social Climate

Compared with residents in proprietary facilities, residents in nonprofit facilities report more cohesion, less conflict, and better organization, as well as more independence and resident influence (Figure 6.3). According to their residents, nonprofit facilities are also more comfortable, a finding that is consistent with their higher level of physical resources. The differences in staff's perceptions of nonprofit and proprietary facilities are smaller than those in residents' perceptions, but staff agree with residents in assessing nonprofit facilities as higher on independence and resident influence. Thus, the work routines and job responsibilities of staff may function as common influences across ownership types and overwhelm the influence of differences in policies and resources between ownership types.

Overall, nonprofit facilities develop a greater sense of community and

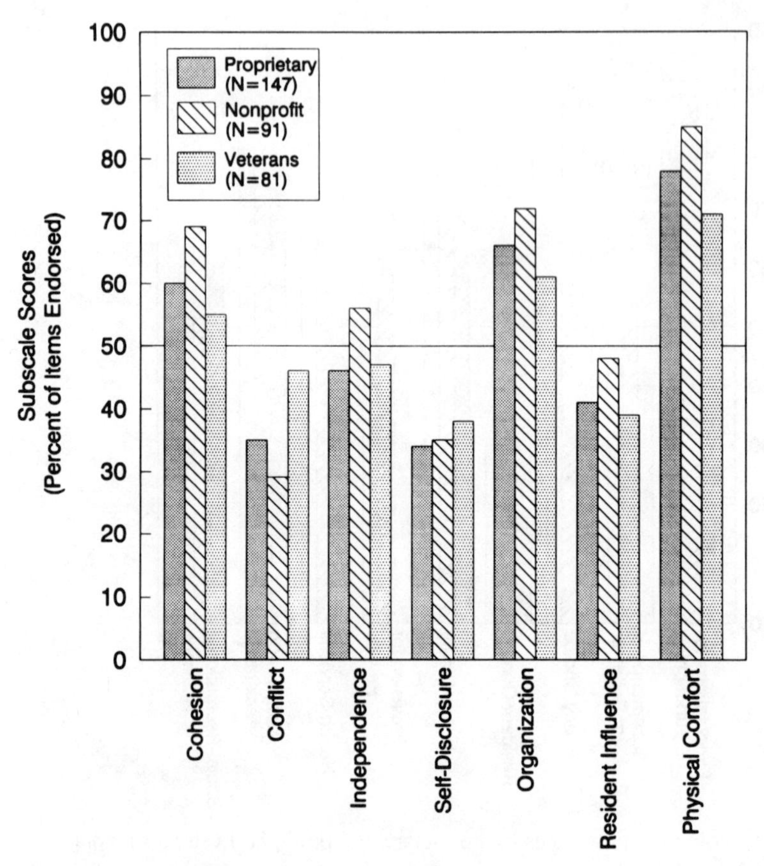

Figure 6.3 Social climates as seen by residents in proprietary, nonprofit, and veterans facilities.

self-direction than do proprietary facilities. These differences may arise in part from the lower turnover among residents in nonprofit facilities and differences in residents' backgrounds. Older people from more privileged backgrounds are more likely to take full advantage of opportunities for personal control and influence. In addition, resident involvement is more consistent with the norms of nonprofit settings where the ethos is one of service to the consumer. More positive interpersonal relationships may also arise from the greater likelihood that residents of nonprofit facilities have preexisting ties to the sponsoring agency and to fellow residents. Some of the differences also may stem from the more flexible policies in nonprofit facilities and the more experienced staff who are more active and supportive of the residents (Barker, Mitteness, & Wood, 1988).

Ownership Differences and Level of Care

The differences between nonprofit and proprietary facilities in physical resources and policies are most striking in the case of residential care facilities. Compared with their proprietary counterparts, nonprofit residential care facilities have more physical resources and give residents more personal control. In contrast, nursing homes and apartments are relatively uniform across the ownership types in their physical resources and policies regarding residents' control. Such uniformity may reflect common influences, such as the impact of governmental regulation or of a shared facility model, which residential care facilities do not have. Residential care facilities operate under less uniform regulation and include several different models, which tend to covary with the type of ownership: Most proprietary residential care facilities are run as community care homes, whereas many nonprofit residential care facilities are found in settings providing a continuum of care.

Nonprofit ownership is associated with better social climate at each of the three levels of care. Among nursing homes, nonprofits have a lower level of conflict, better organization and comfort, and a higher level of resident influence. Among residential care facilities, nonprofit ownership is associated with heightened cohesion, independence, resident influence, and physical comfort. Finally, nonprofit apartments also benefit from greater cohesion, independence, and organization. These differences indicate that the advantages of nonprofit ownership for social climate vary somewhat as a result of the particular regulatory and funding environments in which these facilities exist.

VETERANS VERSUS PROPRIETARY AND NONPROFIT OWNERSHIP

We compared veterans nursing care units and domiciliaries with proprietary and nonprofit facilities after controlling for the level of care. The findings are comparable when we also control for facility size. As shown in Table 6.1, the veterans facilities are intermediate in size between the proprietary and nonprofit facilities. The veterans facilities are housed in older buildings; on average, their construction dates from the early 1940s.

Comparing the three types of ownership, staffing level is highest in veterans facilities, even though the residents are, by a small margin, the most functionally able. The difference in staffing level is greatest for professional staff, such as physicians, psychologists, occupational therapists, and professionally trained nurses. Compared with community facilities, more staff members are men, are over age 50, and have worked in the facility for more than a year. Accordingly, veterans facilities as a group have the highest Staff Resources score, which reflects the diversity and training of staff and low staff turnover.

Resident Characteristics

As noted in chapter 2, community and veterans facilities show marked differences in characteristics of their residents. More than 90% of residents in veterans facilities are men; they are also younger and more likely to be currently married, divorced, or never married (Table 6.1). Such differences in age and marital status suggest that different factors have influenced their move to a veterans facility than those commonly operating for residents of community settings, particularly because people who are married tend not to live in group residences. For married men, moving to a veterans facility may be less stigmatizing than moving to a community setting; relying on outside assistance may be seen as repayment for earlier service, not as a reflection of dependency. For men who are divorced or who never married, factors such as lifelong impairment in functioning or lack of social supports, rather than progressive impairment with advanced age, may dictate the move to congregate housing.

Supporting the view that factors other than frailty and physical dependency contribute to the use of veterans facilities, we found that residents in veterans facilities have somewhat higher functional abilities than do residents in community facilities. For example, compared with community nursing home residents, about 10% to 15% more of the residents of veterans nursing care units are continent, can get in and out of bed by themselves, bathe and dress themselves, go shopping, and handle their own money. Domiciliary residents are also more likely to have these abilities than are residents of residential care facilities, although the differences are somewhat less than for nursing facilities.

The activity pattern for residents of veterans facilities is somewhat different from their counterparts in community facilities. Residents of veterans facilities are more likely to leave the facility for activities in the community but about as likely to participate in activities within the facility. Personal characteristics, such as socioeconomic background, or facility factors, such as a program's encouraging outside activities, may help explain this pattern.

Physical Features

Like nonprofit facilities, veterans facilities provide more physical resources than do proprietary settings. Specifically, veterans facilities have more physical amenities, social–recreational aids, and space for resident functions than do proprietary facilities (Figure 6.1). In contrast, veterans facilities have fewer safety features and are less accessible to community resources than either the proprietary or the nonprofit facilities. These differences, which generally hold for both nursing home (nursing care unit) and residential care (domiciliary) facilities, reflect the age of the buildings that house many of the veterans facilities, the more generous space allowances typically made in older buildings, and their location in full-service medical centers and consequent isolation from community resources.

Policies and Services

As another consequence of their location in full-service medical centers, veterans facilities offer a greater diversity of services than either proprietary or nonprofit facilities, particularly health services and daily living assistance (Figure 6.2). Despite the high level of supportive services, the policies in veterans facilities establish higher expectations for resident functioning than do the policies in community facilities; the differences in expectations parallel the differences in residents' actual functioning.

A number of theorists have argued that there is a necessary trade-off between security and autonomy in a setting (Parmelee & Lawton, 1990). Consistent with the idea that residents may have to relinquish some independence to obtain access to services, the veterans facilities are more structured. Compared with proprietary and nonprofit facilities, veterans facilities give their residents the least flexibility in organizing their daily lives and are least likely to let residents participate in facility governance and to support residents' privacy. However, policies are most clearly articulated and communicated in veterans settings. This high degree of structure probably stems from the centralized management of veterans facilities. For veterans facilities, public ownership does seem related to high levels of service and a more bureaucratic organization.

Social Climate

Despite their somewhat more able residents and their richer staffing and physical features when compared with community facilities, veterans facilities have the least harmonious social climate of the three ownership types (Figure 6.3). Specifically, residents and staff see veterans facilities as highest on conflict and lowest on cohesion, even though residents of veterans facilities, like some residents of nonprofit facilities, share a common affiliation and a preexisting tie to the organization managing the facility.

One factor contributing to the lack of warm personal ties may be the higher proportion of male residents and male staff members in veterans facilities. Community facilities with a higher proportion of male residents tend to be higher on conflict and lower on cohesion; facilities with a higher proportion of male staff are higher on conflict (chapter 7). In a study of psychiatric programs, we found that veterans programs, which had almost all male patients, were lower on involvement, support, and expressiveness than were primarily sex-integrated university hospital programs (Moos, 1974). An additional factor may be the high proportion of never-married and divorced men among residents of veterans facilities; these men may have particular difficulty establishing close relationships.

Veterans and proprietary facilities are comparable in the extent to which residents see the social climate as fostering independence and resident influence; however, veterans facilities are lower than nonprofit facilities in these two areas, according to both staff and resident reports.

The finding that veterans and proprietary facilities place less emphasis on resident influence and independence than do nonprofit facilities is consistent with their relatively restrictive policies on policy choice and resident control. However, the more structured routines, hierarchical decision making, and formal avenues of communication in veterans facilities do not translate into the perception of a social climate that is well organized and orderly. Compared with community facilities, residents and staff see veterans facilities as lower on organization.

Finally, residents' and staff members' perceptions of the physical environments of veterans facilities are discrepant with objectively measured physical resources and with outside observers' perceptions. Veterans facilities have more physical amenities, social–recreational aids, and space availability than do proprietary facilities and are seen by outside observers as being clean and well maintained, yet both residents and staff view the veterans facilities as less pleasant and comfortable than the community facilities. The older buildings and lack of diversity in veterans facilities may help contribute to these perceptions.

In general, our results are consistent with prior research and with intuitive expectations about the differences between ownership types (Elwell, 1984; Vladeck, 1980; Weisbrod, 1989). The results confirm the superiority of nonprofit facilities, particularly in areas that are more difficult to regulate, such as the quality of the social environment. They also confirm that veterans facilities tend to provide high quality care but that they are more regimented and have a less cohesive social climate. Nonprofit ownership is especially strongly related to better resources in residential care facilities, in part perhaps because government regulation is less uniform and the demand exceeds the supply, thus reducing the need for proprietary facilities to respond to market forces.

SIZE AND FACILITY QUALITY

A number of studies have focused on the relationship between facility size and quality. Small facilities are expected to be more homelike and to avoid many of the negative effects of institutions, such as depersonalization and a bureaucratic structure. Large facilities, however, are thought to benefit from some economies of scale and from increased professionalism.

The empirical results are mixed (Davis, 1991). In a study of proprietary nursing homes, Curry and Ratliff (1973) found that residents of smaller facilities had more friends and more social interactions within the home. In contrast, Weihl (1981) reported that among residents of homes for functionally independent older adults in Israel, satisfaction with the social milieu, existence of friendships, and feelings of belonging were higher in homes with more residents. In a survey of federally assisted housing, Lawton, Nahemow, and Teaff (1975) found that the number of units in the facility was not related to friendships within the facility, activity participation, or morale.

Some researchers have found that larger nursing homes tend to give higher quality care as reflected in staffing levels and expenditures (Greene & Monahan, 1981) and in treatment resources (Kosberg & Tobin, 1972).

In contrast, Greenwald and Linn (1971) reported that staffing was lower in larger nursing homes, which also received lower ratings in such areas as cleanliness and activity, patient satisfaction, housekeeping, and proper diet. Meiners (1982) found that larger nursing homes had slightly lower costs, but that economies of scale were quite small (see also Caswell & Cleverley, 1983; Lee & Birnbaum, 1983). A number of researchers have found a curvilinear relationship between size and costs, with medium-sized facilities being the most efficient (McKay, 1991; Ullmann, 1985).

Facility Size and Resources

We focus here on the associations between size (number of residents) and physical features, policies and services, and social climate in the community facilities. To identify the correlates of size, we calculated partial correlations controlling for level of care. Because size and ownership are related (nonprofit facilities are larger), we also controlled for the differences between ownership types. In addition, we examine how the associations vary in proprietary and nonprofit facilities and comment on some findings in the veterans facilities.

Physical Features

In the case of community facilities, larger facilities enjoy a number of advantages in terms of their physical features: more physical amenities, social–recreational aids, prosthetic and orientational aids, safety features, and space for staff functions. On the other hand, they are less spacious.

These relationships are somewhat stronger in proprietary than in nonprofit facilities. As Figure 6.4 illustrates, the number of physical amenities increases more strongly with size in proprietary than in nonprofit facilities. A similar pattern also applies to social-recreational aids, prosthetic aids, safety features, and staff facilities. Thus, economies of size for expenditures on the physical plant appear to be more important in proprietary than in nonprofit facilities.

Nonprofit facilities may be cushioned from the full force of economic pressures because they do not need to make a profit for investors. In addition, nonprofit facilities are more likely to be part of a large, multilevel facility; accordingly, small units can share some resources with the larger organization and achieve economies of scale in that manner. In fact, for proprietary facilities, economies of size are more evident in facilities that are owned by individuals or small corporations than in facilities that are part of large corporate chains. These findings are consistent with research that finds different cost functions for different ownership types (e.g., Arling, Nordquist, & Capitman, 1987; McKay, 1991).

As in the community sample, larger veterans units have more space for staff but less for residents. Otherwise, size is not related to physical features in veterans facilities, probably because of the centralized adminis-

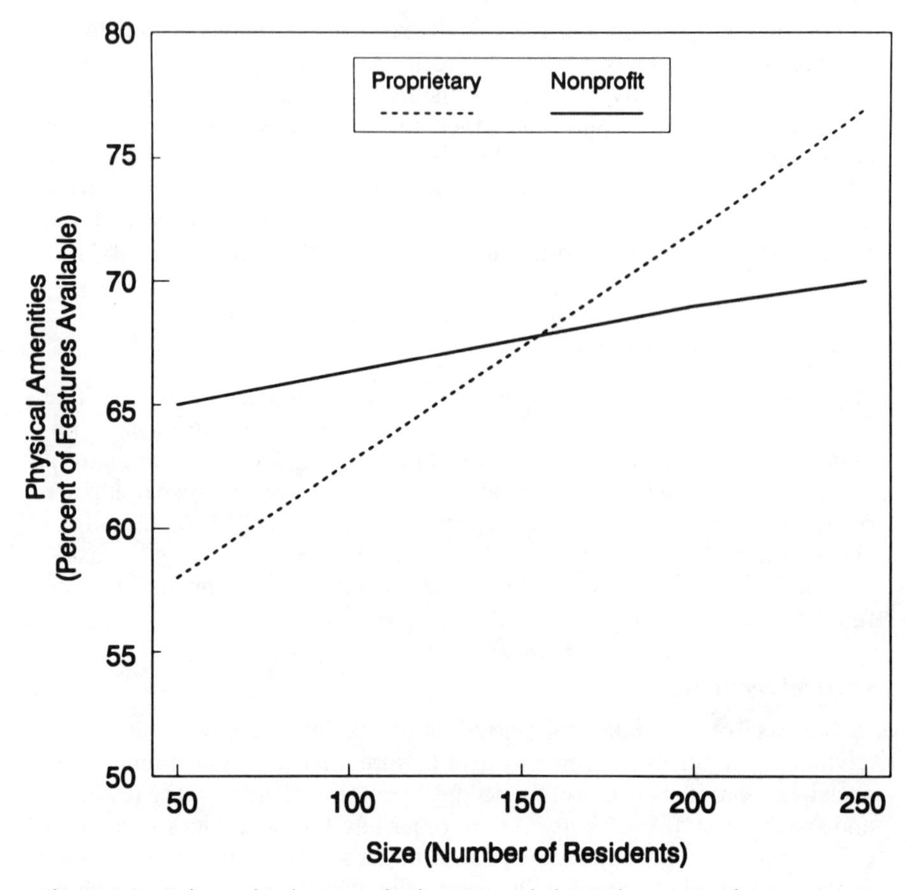

Figure 6.4 Relationship between facility size and physical amenities for proprietary and nonprofit facilities.

tration of the Department of Veterans Affairs and the common policies and regulations guiding individual units. In addition, veterans facilities, like nonprofit ones, are buffered from some direct economic pressures.

Policies That Promote Personal Control

Overall, larger community facilities have policies that provide more opportunities for personal control; specifically, they have more choice in activities of daily living, more resident control, and more privacy. In addition, larger facilities are more likely to institute procedures for systematically communicating policies to residents and staff. Similarly, larger veterans facilities make more formal provision for resident involvement in decision making. The relationships between size and facility policies are fairly constant across ownership types.

These findings run counter to the idea that large size necessarily leads to a more restrictive and less individualized environment. Larger facilities are more likely to have at least a minimum number of interested and functionally able residents who participate actively in decision making and who are likely to have more input into facility governance. Perhaps because larger facilities cannot rely on informal communication and have sufficient numbers of staff and residents as potential participants, these facilities are more likely to have formal orientation programs, regular staff meetings, and a newsletter. The presence of a minimum number of able residents and professionally trained staff may also encourage management to formalize opportunities for resident choice. However, formalized procedures may contribute to making large facilities somewhat more bureaucratic than small ones.

Services

Larger community facilities offer more health services and social–recreational activities and have a more diverse and experienced staff. However, the ratio of staff to residents is slightly lower in large facilities than in small ones, particularly for staff not involved in patient care activities. Thus, larger facilities can reduce personnel costs in maintenance and administration areas where the workload does not appear to increase in proportion to the number of residents.

As the patient population increases and facilities hire more staff in order to maintain a fixed ratio of patient-care staff, the additional staff members tend to fill more diverse roles. Larger facilities are more likely to offer more specialized services and activities geared to the needs and interests of a minority of residents. For example, a facility with 30 residents is unlikely to offer an activity that appeals to only 10% of the residents. However, in a facility with 150 residents, the same proportion of interested residents is likely to justify offering the activity. Evidence of this pattern is that the use of services and activities (the percentage of residents using them in a typical week) declines as facility size increases.

Size is not as closely related to services in veterans facilities, although larger veterans facilities do have more social–recreational activities. (For discussion of how the effects of size vary in facilities that offer different levels of care, see Lemke & Moos, 1989a.)

Social Climate

As we have seen, larger facilities tend to be richer in physical features and services and less restrictive in their policies. Nonetheless, large institutions may be more impersonal, discordant, and chaotic due to their somewhat more crowded conditions and their more complex social structure. Consistent with these ideas, residents and staff in larger community facilities report more conflict and somewhat poorer organization. Thus, large facility size appears to reduce efficiency and to cause more strain in relationships. In the view of residents, larger veterans units also show a higher level of conflict than do small ones.

Large facility size may reduce pressure to conform and to suppress anger. Greater privacy and a broader range of possible friendships may lead residents in larger facilities to express some feelings more openly. We found as evidence that residents also report more self-disclosure in larger facilities. Greater size may also contribute to the frequency of situations causing dissatisfaction and anger and may increase the likelihood that a resident or staff member will witness expressions of conflict by others. However, residents and staff report as much cohesion in large as in smaller facilities, a finding that suggests that supportive relationships are not more difficult to establish in large facilities.

The sheer number of residents and staff and the increased hierarchy of large institutions may make communication more difficult, but these factors may also encourage a greater emphasis on having residents do as much as they can for themselves. Thus, in a large facility, staff may be less able to give personal attention to residents, residents may have somewhat more freedom because of their greater anonymity, and residents may be able to assume more differentiated roles within the system. In fact, residents and staff report more independence in larger facilities. This finding is consistent with the fact that larger facilities provide their residents with more privacy, options for organizing their daily activities, and opportunities for participating in facility governance.

Economy of Scale and Organizational Complexity

Previous research has generated some evidence for economies of scale in the operation of nursing homes (Greene & Monahan, 1981; McKay, 1991; Meiners, 1982; Ullmann, 1986). Economies of scale allow a larger number of residents to bear the costs of physical and staff resources; accordingly, per-person costs are lower and larger facilities can provide a wider array of services for a given expenditure. Consistent with these ideas, we found that larger facilities have a more diverse and experienced staff, provide more services, and have more physical resources. We also found that larger facilities tend to operate with a lower staffing level in non-patient care roles, a practice that can result in lower personnel costs.

Size not only influences the economics of residential facilities, it may also have an impact on quality of care by means of social structures and relationships. Larger organizations tend to develop more complex and hierarchical structures with more differentiated roles. We found evidence that large facilities evolve more formal arrangements for handling particular functions, such as communicating policies and expectations to staff and residents and obtaining resident input. However, larger facilities also tend to be somewhat more chaotic and to have a higher level of conflict, in spite of their efforts to ensure clear communication.

Overall, larger proprietary facilities seem to use their resources to enhance the physical environment and to provide more services. Size is less important for nonprofit facilities, although they also tend to provide more services as size increases. The negative impact of large size seems to come mainly in increased conflict and confusion.

STRUCTURAL VARIABLES AND FACILITY QUALITY

Structural variables such as ownership, size, and level of care provide a general context within which facility quality develops. These factors influence the selection of residents and staff, the regulatory and economic environment within which the facility functions, and the organizational ethos that develops.

In earlier chapters, we described the importance of facility type as a determinant of the resources available to residents. Nursing homes are oriented toward providing a self-contained environment that facilitates the provision of care and services. The emphasis is on the public, communal facet of residents' lives, with minimal accommodation of residents' independence and privacy. Residential care facilities also provide a level of support and services that tends to restrict residents' autonomy and independence, but these facilities have fewer public, communal features than nursing homes.

Apartments have a two-tiered structure. At the facility level, the shared, public aspects of life tend to be explicitly recognized in spite of the minimal level of supportive services provided. At the individual apartment level, the individual's privacy and independence are emphasized. Overall, each facility type represents a different compromise between the provision of security and communality and an orientation toward autonomy and separateness (Parmelee & Lawton, 1990).

Similarly, the three ownership types we examined here show somewhat different patterns of resource allocation. The comparison of veterans and proprietary facilities supports the view that there is a trade-off between environmental support and resident autonomy within a setting. Although veterans facilities provide more physical and staff resources than do proprietary facilities, they are also more restrictive. The pattern in nonprofit facilities, however, shows that there need not be such a direct trade-off; many facilities could make some improvement in each of these areas without any accompanying loss in the other. Larger size also appears to allow a facility to offer both a more supportive environment and one in which residents can exercise more control.

Although factors such as level of care, ownership, and size influence the resources and social climate that develop in a facility, these factors do not operate as rigid constraints. Other factors, such as financing and government regulation, also have an impact on resources and social climate.

For example, although proprietary facilities have fewer supportive features than do nonprofit facilities, no such difference is found for nursing homes, which are generally constructed within stricter architectural guidelines. Small facilities are able to provide fewer physical and staff resources for their residents, but the relationship between small size and these disadvantages is weaker in nonprofit or veterans facilities. Accordingly, the potential problems that may be associated with facility ownership and size can be overcome by careful planning and a system of ongoing monitoring and feedback of information to program managers and staff.

III

APPLICATIONS FOR PROGRAM EVALUATION

7

Developing Harmonious, Resident-Directed, and Well-Organized Social Climates

In part II, we presented descriptive information about residential programs. Practitioners and researchers can use such information as process and structural indices of quality of care. In part III we turn to outcome indices of quality and focus on how facility features influence the lives of residents and staff. Chapter 7 examines the relationship between program features and a facility's social climate. In chapters 8 and 9, we examine the impact of program factors on residents; we include facility-level analyses (chapter 8) and individual-level analyses (chapter 9). In chapter 10, we illustrate how the MEAP can be used to help understand group and individual functioning over time.

The finding of considerable variability in the social climates of group residences (chapter 5) raises the question of why social climates develop in such disparate ways. In chapter 6, we examined the impact of institutional factors, such as ownership and size, on social climate; in chapter 7, we examine the influence of the three other sets of dimensions assessed by the MEAP, namely, the aggregate characteristics of residents and staff, physical features, and policies and services.

A MODEL OF THE DETERMINANTS OF SOCIAL CLIMATE

To address the issues of the determinants of social climate, we expanded the conceptual framework depicted in Figure 1.1 to show the connections between panels I and III in that model. As shown in Figure 7.1 (see also Moos & Igra, 1980), the institutional context (factors such as level of care,

This chapter is coauthored by Christine Timko.

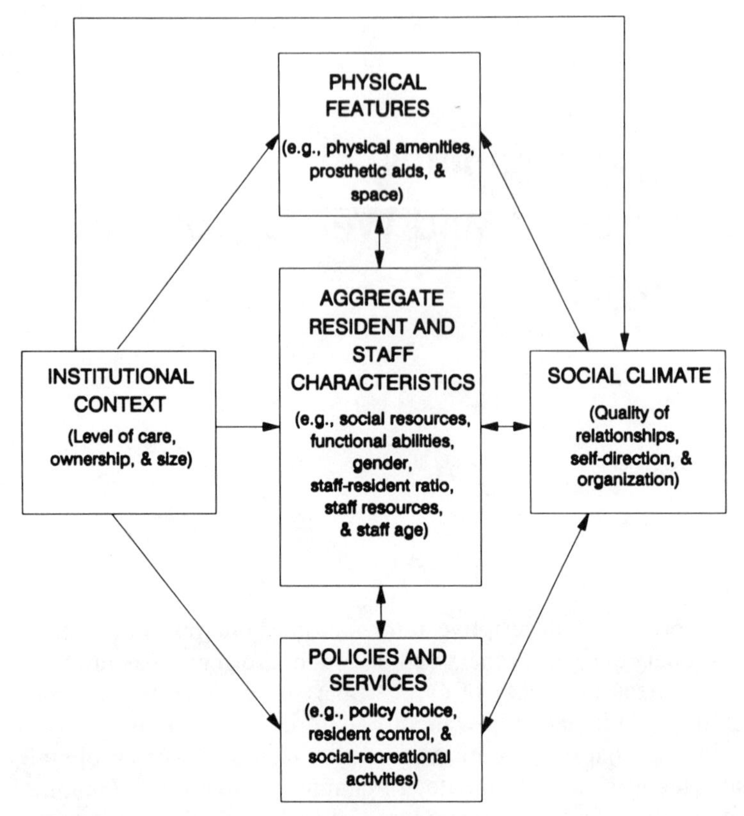

Figure 7.1 A model of the determinants of social climate in group residential facilities.

ownership, and size) and the three sets of environmental factors can directly or indirectly influence social climate. For example, residents with more social and functional resources may engage in more activities and social interaction and thus promote cohesion and a sense of independence. Social–recreational aids may enhance cohesion by encouraging joint activities; policies that allow residents more choice may contribute to resident independence. Residents with more social and functional resources may also be more likely to request and be given a voice in facility decisions; in turn, policies that give residents control may contribute to the perception of high resident influence.

We use the framework and some of our earlier findings (chapters 3, 4, and 5) to focus on the associations between social climate and specific aggregate resident and staff characteristics, physical features, and policies and services in the sample of 262 community facilities. The aggregate resident characteristics are the residents' social resources, functional abilities, and gender composition. The staff characteristics are the staff–resident ratio, staff resources, gender composition, and age. Within the domain of physical features, we look at how physical amenities and social–recreational aids, prosthetic aids and safety features, and the space available for resident and staff functions affect social climate. Finally, we examine the impact of policy choice, resident control, privacy, policy clarity, and facility-planned social–recreational activities on the social climate.

We focus on five of the seven dimensions of social climate measured by the Sheltered Care Environment Scale (SCES): the quality of relationships (Cohesion and Conflict), the emphasis on resident self-direction (Independence and Resident Influence), and orderliness and clarity (Organization). To identify the predictors of these social climate dimensions, we calculated partial correlations (controlling for level of care) between facility characteristics and residents' perceptions of the social climate. These partial correlations reflect the associations between the three sets of facility characteristics and social climate, independent of the facility's level of care.

As noted in chapter 5, residents and staff make equally reliable judgments of the social climate, although within different contexts. Here we emphasize the residents' view of the facility. Although the residents' well-being is a primary concern of planners and administrators, they need to develop effective social climates for residents without having a negative impact on staff. Accordingly, in this chapter, we note some consistencies in the determinants of residents' and staff's views. With few exceptions (Timko & Moos, 1990), the aggregate resident and staff characteristics, physical features, and policies and services that are related to residents' views of social climate are related in similar ways to staff's views.

DETERMINANTS OF SOCIAL CLIMATE

Institutional Context and Social Climate

Earlier chapters described the relationships of level of care (chapter 5) and ownership and size (chapter 6) to social climate. Consistent with our model, we found that these three aspects of the institutional context influence social climate.

Gerontologists have described nursing homes as total institutions, lacking privacy and requiring conformity. They argue that these features tend to increase residents' dissatisfaction with peers and staff as well as their dependency on staff. The high level of impairment of nursing home residents and the resulting care routines may also make it more difficult for nursing homes to function smoothly and efficiently. In contrast, independent housing is likely to allow older people greater freedom to find their desired mix of privacy and support and to meet their varied preferences for a daily routine. Consistent with this perspective, residents see more conflict, less independence, and poorer organization in nursing homes than in congregate apartments. Residential care facilities are like nursing homes in their level of independence and like apartments in their levels of conflict and organization (see Figures 5.1 and 5.2).

Consistent with theoretical expectations that nonprofit facilities provide a superior quality of care, we found that nonprofit facilities develop a greater sense of community and resident self-direction than do proprietary facilities. Larger size is associated with residents' and staff's assessments of more independence, more interpersonal conflict, and poorer organization (chapter 6).

We believe that these findings are due in part to variations in the aggregate characteristics of residents and staff, physical features, and policies and services associated with level of care, ownership, and size. Thus, the greater cohesion and resident self-direction in nonprofit facilities may be a function of their richer physical features, more flexible policies, more privileged residents, and more experienced staff. Larger facilities may be higher on independence because they are more likely to have policies that enhance residents' personal control; they are also somewhat more crowded, which may contribute to a higher level of conflict.

Resident Characteristics and Social Climate

Residents' abilities, personal resources, and typical patterns of behavior help determine the social climate that develops in a congregate residence. In general, we found that facilities with more women and more socially privileged residents develop more positive social climates.

Residents' Social Resources

In facilities with higher aggregate social resources, residents report greater cohesion and less conflict (Table 7.1); the findings were comparable for staff. Several factors may contribute to the greater harmony in settings where residents have more social resources. Less-privileged resident groups may experience more stress because their facilities have fewer resources, such as privacy and space. Depending on their social backgrounds, residents may have different views of the open expression of conflict; for example, among socially advantaged residents, norms may encourage the suppression of anger. In addition, residents with fewer social resources may be less able to cope with the demands inherent in a group living situation and may react with complaining and angry outbursts. They may also find themselves less able to maintain equal status with staff and to reciprocate staff social interactions; this inability may, in turn, lead to lower levels of cohesion.

According to residents (Table 7.1) and staff, facilities that house residents with more social resources place more emphasis on independence and give residents more voice in the facility. As with the findings on conflict, these findings may reflect different norms related to residents' social status. Based on their past experience, residents with more social re-

Table 7.1 Partial correlations between aggregate resident and staff characteristics and social climate as assessed by residents

Resident and staff characteristics	Social climate				
	Cohesion	Conflict	Independence	Resident Influence	Organization
RESIDENTS					
Social Resources	.16*	−.39***	.20**	.26***	.24***
Functional Ability	.11	−.06	.06	.16**	.15*
Women residents (%)	.17**	−.19**	.17**	.05	.13*
STAFF					
Staff–resident ratio	.02	−.09	−.15*	−.08	.04
Staff Resources	.26***	−.07	.20**	.25***	.14*
Women staff (%)	.05	−.17**	−.12*	.00	.02
Staff over age 51 (%)	.06	−.08	−.01	.07	.21***

$N = 262$ facilities; partial correlations control for level of care.
*$p < .05$; **$p < .01$; ***$p < .001$.

sources may be more responsive to staff encouragement of independence and more effective in voicing their opinions and concerns. If staff see that residents value and effectively use these opportunities, staff may be more likely to institute and maintain procedures that accommodate resident independence and influence. In addition, although social status may decline somewhat with age, older people with higher social status may retain relatively more power in their relationships with staff members than do residents with lower social status.

Social privilege is also related to better organization as perceived by residents (Table 7.1). In part, this relationship is indirect; residents with higher social status tend to live in smaller facilities employing older, more experienced staff, and these attributes may contribute to better organization. In addition, resident groups with more resources may have a broader array of alternative sources of support; as a consequence, the overall demand on staff may be less and fewer crisis situations may arise.

Residents' Functional Abilities

Although residents' functional abilities are often perceived as a major variable constraining the social climate that develops within a facility, our findings indicate that, once level of care is controlled, resident functioning is not strongly related to facility social climate. Residents in facilities with higher aggregate functioning report slightly higher resident influence and better organization. As with social resources, higher functioning resident groups may be better able to take advantage of opportunities for control and may reduce the level of demand on staff. Although some researchers have suggested that competent residents are likely to create additional problems for staff by being more demanding and hard to satisfy (Buffum & Konick, 1982), our findings do not support this view.

Relatively impaired resident groups appear about as likely to develop harmonious relationships as do groups with more-intact functioning. Similarly, in the views of both residents (Table 7.1) and staff, independence is not related to residents' actual level of functional dependency. In other words, a climate that supports resident autonomy can develop even in a setting with a very impaired, physically dependent resident population.

Proportion of Women Residents

According to residents and staff, facilities with a higher proportion of women residents are higher on cohesion and organization; residents also see them as lower on conflict (Table 7.1). These findings are consistent with the fact that veterans facilities, which are populated primarily by men, tend to be less cohesive and orderly and higher on conflict than are community facilities (chapter 6). Women may be more oriented toward social relationships than are men. In addition, facilities with a higher proportion of men residents, such as board-and-care homes or urban apartments, may draw from a different cross section of the elderly population than do those with a more representative proportion of women residents.

Although the older people who are currently living in group housing were generally socialized into gender roles in which men are seen as more independent than women, groups with more women residents place more emphasis on resident self-direction. Residents see living groups with a higher proportion of women residents as higher on independence; staff view them as higher on resident influence. Because older women are more likely than older men to have organized their own household and to have provided care to others, women residents may resist being cared for and having their daily routine dictated by others; in a residential setting, they may elicit a greater emphasis on independence and on resident influence.

Staff Characteristics and Social Climate

Higher staffing level has been used as an indicator of better quality care and might be expected to be related to the development of cohesive, well-organized social climates with an emphasis on independence. Staff training and experience also are expected to influence a facility's social climate. Research has indicated that better-educated staff have more positive attitudes toward older people than do less educated staff (Chandler, Rachal, & Kazelskis, 1986); older staff have more positive attitudes than do younger staff (Wright, 1988). Moreover, staff members who have more positive attitudes toward residents as a group see them as preferring to be independent in their interactions with others (Caporael, Lukaszewski, &

Culbertson, 1983). In general, we found that background characteristics of staff are related to facility social climate and that the staffing level is not.

Staffing Level and Staff Resources

After controlling for level of care, higher staffing has little relationship with social climate in the residential facilities we studied: Higher staffing is not associated with more cohesion or better organization in the facility; in fact, it is associated with slightly less independence as reported by residents. With more staff time available per resident, staff may inhibit residents from completing activities independently (Avorn & Langer, 1982). Residents in such settings may have fewer opportunities to practice self-care skills and may come to believe that they are incapable of managing without help (Baltes & Reisenzein, 1986).

In contrast with staffing level, staff training and experience seem to promote a harmonious, independence-oriented, well-organized environment (Table 7.1). In facilities with higher Staff Resources, residents report greater cohesion, independence, resident influence, and organization; staff report slightly higher cohesion. Staff with more training and longer tenure seem to be more successful in creating a cohesive and well-organized program for residents, perhaps because the staff's experience enables them to manage effectively and communicate clearly. Although most of the care delivered to residents is by employees who have little formal training (Johnson & Grant, 1985), the presence of well-trained professionals in leadership roles may have a major impact on the quality of care.

Proportion of Women and Age of Staff

Generalizing from our results on the gender composition of the resident population, we might expect that living groups with a higher proportion of women staff members would emphasize interpersonal relationships and resident self-direction. Like facilities with more women residents, those with more women staff members are viewed by residents as having less conflict. However, facilities with a higher proportion of women staff members are also seen as placing less emphasis on independence. Residents may view women in staff roles as more nurturing than men and as less inclined to encourage residents' independence.

We also looked at the proportion of staff over age 50. The proportion of older employees did not affect residents' perceptions of relationships or of the emphasis on self-direction, but staff report less conflict in settings where more of the staff are in this upper age range. In addition, according to both residents and staff, a higher proportion of older employees contributes to better facility organization. A more mature staff, who are likely to have more experience working in this field, appears to contribute to predictability and efficiency in the facility.

Physical Features and Social Climate

We focus here on the associations between social climate and physical features that enhance resident comfort (physical amenities and social–recreational aids), provide a secure environment (prosthetic aids and safety features), and provide space for resident and staff functions (space availability and staff facilities). In general, these physical resources help promote a cohesive, resident-directed social climate.

Physical Amenities and Social–Recreational Aids

Residents report more cohesion, independence, and resident influence in living groups with more physical amenities and social–recreational aids (Table 7.2); the findings were comparable for staff. These physical features may enhance cohesion and resident autonomy because they encourage residents to leave their private quarters to use common areas, and they provide resources with which residents can initiate activities, engage in social interaction, and make productive use of their time. These findings show that physical features designed to improve pleasantness and to facilitate social interaction can contribute to resident autonomy.

Prosthetic Aids and Safety Features

Residents in settings with more prosthetic aids and safety features tend to report greater independence (Table 7.2). Prosthetic features, such as those that make the facility more accessible to residents with mobility impairments, may encourage full use of the environment and diminish the need to rely on others for assistance. Safety features, such as call buttons, nonslip surfaces, and staff monitoring of public areas, may similarly encourage residents to use all areas of the building and reduce their need to seek staff assistance. Thus, supportive features do not seem to increase

Table 7.2 Partial correlations between selected physical features and social climate as assessed by residents

	Social climate				
Physical features	Cohesion	Conflict	Independence	Resident Influence	Organization
Physical Amenities	.28***	−.07	.18**	.23***	.06
Social–Recreational Aids	.30***	−.02	.27***	.23***	.06
Prosthetic Aids	.06	.11	.15*	.11	−.13*
Safety Features	.19**	−.05	.20**	.00	.15*
Space Availability	.21**	−.18**	.04	.18**	.16**
Staff Facilities	.31***	.00	.25***	.17**	.03

$N = 262$ facilities; partial correlations control for level of care.
$^*p < .05$; $^{**}p < .01$; $^{***}p < .001$.

dependency by reducing environmental challenge, nor do they seem to carry the message that residents are viewed as not competent.

Safety features are also related to more cohesion and better organization. Safety features may increase group interaction and communication by encouraging residents to make full use of public areas. In contrast, more prosthetic aids are associated with reports of poorer organization. Facilities with more prosthetic aids tend to be larger, and their size and more institutional orientation may contribute to confusion.

Space for Resident and Staff Functions

Little research has been done on the effects of crowding in residential settings for older people, but work with other age groups shows that crowded conditions are related to impaired social relationships and poorer mental and physical health (Baum & Paulus, 1987). Consistent with these findings, Christensen and Cranz (1987) noted that in age-segregated public housing, residents report more socializing among residents in facilities with larger lobbies.

Residents report more cohesion, less conflict, more resident influence, and better organization in facilities with more space reserved for resident functions (Table 7.2). Staff also perceive less conflict, higher resident influence, and better organization in more spacious facilities. In congregate residences, the availability of separate areas and adequate space for diverse functions may not only encourage relationships by giving residents the opportunity to join others in preferred activities but also reduce potential sources of conflict. Adequate space may reduce the tendency to restrict residents through rules and regulations and enable residents to regulate their interpersonal contact, thus giving them a greater feeling of control. Adequate space also facilitates planning and conducting activities, thus contributing to the facility's efficient management.

Where more space is available for staff, residents view the social climate as more cohesive (Table 7.2). Residents see staff as being more available and supportive, in spite of the fact that staff may spend less time in actual proximity to residents. By providing a respite from caregiving responsibilities, such features may improve staff morale and allow them to be more supportive in their interactions with residents. In contrast, staff report more conflict in facilities where they have more space; perhaps when separate staff areas are available, staff are more likely to talk about their conflicts in dealing with residents and other staff.

Where staff have more space, residents view the social climate as higher on independence and resident influence (Table 7.2); staff view it as higher on resident influence. When staff members are not in constant physical proximity to residents, they may do more to encourage resident independence and may interfere less with residents' activities. Having space of their own, staff may be more willing to cede control of other areas of the facility to residents.

Finally, staff see living groups with more staff facilities as somewhat less well organized. Facilities with more staff facilities also tend to be larger, and their size may contribute to perceived disorganization and conflict. Isolation in separate staff areas and hierarchical organization of large facilities may contribute to staff perceptions of disorder.

Policies and Services and Social Climate

In this area, we focus on the associations between indices of institutional freedom and control (policy choice, resident control, provision for privacy, and policy clarity) and facility-planned activities and social climate. Prior studies show that policies that give older residents more choice and control are associated with more resident interaction and activity. In addition, residents who live in facilities that encourage resident choice and control see themselves as more independent and as having a greater voice in decision making (Rodin et al., 1985). Privacy is another highly salient feature in group settings (e.g., Berdes, 1987) and can further contribute to opportunities to exercise choice and control. Policy clarity may increase predictability for facility residents and staff and, consequently, their feeling of control; it may also contribute to more efficient facility operation. Finally, we expect that a diverse program of facility-planned activities may create opportunities for meaningful social interaction.

Policy Choice, Resident Control, and Privacy

According to residents, living groups with more policy choice, resident control, and privacy are higher on independence and resident influence (Table 7.3). Staff also see more resident self-direction in these facilities. Thus, policies that allow residents to determine their daily routine and that establish formal avenues for residents to affect the facility's operation appear to increase residents' independence and influence.

Table 7.3 Partial correlations between selected policies and services and social climate as assessed by residents

	Social climate				
Policies and services	Cohesion	Conflict	Independence	Resident Influence	Organization
Policy Choice	.07	−.09	.21**	.13*	.04
Resident Control	.20**	−.07	.32***	.31***	.07
Provision for Privacy	.24***	−.22***	.35***	.21***	.17**
Policy Clarity	.15*	.02	.26***	.20**	.01
Social–Recreational Activities	.17**	−.01	.23***	.10	−.01

$N = 262$ facilities; partial correlations control for level of care.
$^*p < .05$; $^{**}p < .01$; $^{***}p < .001$.

Residents in facilities with more resident control and privacy also report greater cohesion. Thus, a residents' council may serve a relationship as well as a task function; that is, participation may strengthen relationships among residents and staff as well as give residents a voice in decisions. In addition, residents may be more likely to remain socially engaged if they feel they exercise some control over their environment and can withdraw from unwanted social contact. In this vein, researchers have noted that enforced contact and lack of privacy may actually reduce social interaction. When individuals do not have adequate privacy, they may try to maintain personal boundaries by engaging in behavior that isolates them from positive interactions as well as from unwanted intrusion (Baum & Paulus, 1987). The ability to withdraw from social contact may help reduce conflict as well. Residents report less conflict in facilities with more provisions for privacy (Table 7.3).

Staff members sometimes express the fear that individualizing care routines will lead to confusion and inefficiency and that residents who exercise autonomy may disrupt orderly routines and procedures. However, our findings show that high choice and control need not lead to disorganization. Neither residents (Table 7.3) nor staff report poorer organization in facilities whose policies emphasize resident choice, control, and privacy. In fact, privacy is associated with better organization as assessed by residents. For residents, privacy may reduce the level of confusion by allowing them to withdraw from the communal domain when they choose to do so. Thus, high levels of choice, control, and privacy can be provided without reducing the level of efficiency or increasing confusion for residents or staff members.

Clarity of Policies

Residents report more cohesion in facilities with formal avenues for communicating policies to residents and staff (Table 7.3); these facility policies are not related to staff's perceptions of cohesion or conflict. Formal arrangements may increase group cohesion by clarifying expectations and reducing uncertainty for residents. As with formal provision for resident control, residents may also use formal mechanisms for relationship as well as for task functions. For example, an orientation program for new residents offers an opportunity to get acquainted as well as to learn the rules and routines.

We expected and found that clearly stated policies contribute to self-direction among residents as seen by both residents (Table 7.3) and staff, probably by making the environment more predictable. Ambiguous policies may contribute to feelings of helplessness.

Contrary to expectations, formal means of communicating policies do not seem to contribute to either residents' or staff members' assessments of organization. Facilities with informal means of transmitting expectations are equally able to achieve orderly operation as those with formal structures.

Social–Recreational Activities

Residents judge living groups with more social–recreational activities as more cohesive and as offering more independence; staff judge these living groups as higher on independence and resident influence. The provision of social–recreational activities may contribute to harmonious relationships by facilitating residents' participation in shared activities. Facility-sponsored social–recreational activities also foster a climate of independence and do not appear to undermine residents' feelings of competence and autonomy. The availability of a broad range of activities may permit residents to choose desired activities and, thus, to see themselves as active and exercising some control.

Variance Explained by Sets of Predictors

To integrate the foregoing material with the conceptual framework presented in Figure 7.1, we look at how the institutional context and the sets of aggregate resident and staff factors, physical features, and policies and services influence social climate. Specifically, we conducted multiple regression analyses to identify the unique and common variance explained by the four sets of predictors for each of the social climate dimensions. The findings generally confirm the model of the determinants of social climate and show that the influence of the institutional context (level of care, ownership, and size) on social climate is largely shared with the three other sets of predictors.

Taken together, the four sets of predictors explain a substantial amount of the variance in residents' perceptions of social climate. The variance explained ranges from 23% for cohesion to 42% for independence (mean = 31%). In general, after the institutional context is considered, each of the three other sets of factors adds significant incremental variance to residents' reports of social climate dimensions. With some exceptions, the basic findings for staff are comparable.

The institutional context independently explains relatively little of the variance in social climate. There is no independent association between level of care and social climate. Ownership and size show small but significant associations with social climate (accounting for 1% to 3% of the variance) even when the three other sets of determinants are controlled. Nonprofit facilities are higher on cohesion, independence, and resident influence; larger facilities have more conflict and poorer organization.

In general, however, social climate is a joint outgrowth of two or more sets of determinants; more than half of the predictable variance in the social climate dimensions is shared between two or more sets of determinants. Overall, the findings show that administrators need to consider combinations of these factors when they try to change the social climate in their facility.

INTERVENTIONS TO IMPROVE SOCIAL CLIMATE

A number of theorists have developed guidelines to help facility administrators and staff understand and change the social climate likely to emerge in a residential facility (e.g., Berdes, 1987; Bowker, 1982). The relationships we have found may help identify group residences where interventions can contribute to a more harmonious, resident-directed, and well-organized social climate than might otherwise develop.

Developing a Harmonious Climate

The communal quality of congregate residential settings has both potential benefits and potential drawbacks. It offers the possibility of a supportive community as individuals become less able to sustain relationships over physical distance. By the same token, the enforced contacts in such settings may give rise to stress and conflict in interpersonal relationships. An important goal for facility administrators is to establish a cohesive social climate in which conflict is low to moderate.

Managers and staff need to be aware of the interpersonal problems likely to arise among specific groups of residents. For example, facilities drawing their resident population from less socially privileged strata or serving more men are likely to develop less harmonious social climates. Women and those who have more social resources are likely to establish a social climate that is more cohesive and in which conflict is low. Staff who are aware of these tendencies can counteract them with efforts aimed at reducing alienation and disharmony.

One important way to enhance the quality of interpersonal engagement is to support the use of the building's public areas. An environment that is attractive, has features that foster social behavior and recreational activities, is safe, and offers adequate communal space encourages residents to use public areas and to be involved in positive interactions. The program can contribute to active involvement by providing a broad range of facility-planned activities in these public areas and by offering opportunities for residents to meet for the purpose of jointly influencing facility policies.

Along these lines, Bakos and his coworkers (1980) implemented an environmental management process by which residents and staff jointly planned changes to improve the design and increase social interaction in a geriatric treatment facility. Consistent with our ideas on the determinants of social climate, the changes focused on both physical features and activities. The physical changes involved rearranging the dayroom furniture and constructing modular activity centers by introducing movable partitions and adding a kitchenette and snack area. The television set was placed in one corner of the dayroom where it would not be the central focus. New social activities, such as a cooking class, were initiated. As expected, residents and staff spent more time in the dayroom. In addition, residents' activity levels increased and their behavior became more functional and adaptive. Joint resident and staff participation undoubtedly helped to make this project a success. In fact, residents who participated in the decision-making process showed the most improvement.

Another key way to maintain residents' willingness to interact and to reduce conflict is to provide ample privacy. At least in the United States, opportunities to act without scrutiny by others and to withdraw from contact appear vital to sustaining warm, supportive relationships and to reducing conflict. Residents also feel closer in settings where the physical environment allows staff to withdraw from direct contact with residents. Thus, in facilities with little interpersonal warmth and much conflict, administrators can indirectly influence the social climate by increasing residents' sense of control and privacy.

Supporting Resident Self-Direction

A significant body of research supports the view that older residents in congregate settings benefit from an emphasis on independence and from opportunities to exercise control. Like the quality of relationships, resident self-direction depends in part on predispositions that residents bring to the setting. Thus, for example, settings that draw their residents from the more socially advantaged tend to place a greater emphasis on resident autonomy. However, as we have seen, administrators can encourage independence even when residents are frail and in need of high levels of staff assistance (see also Clark, 1988).

The physical environment also can contribute to resident self-direction. A physical environment that supports resident activity, engagement, and control by way of amenities and social–recreational aids can help communicate to residents that they are viewed as autonomous. An accessible and safe environment contributes to the emphasis on independence, and the availability of space for resident functions is related to their level of perceived influence.

As expected, policies that provide residents more choice in daily routines, formal input in decisions, and privacy have a positive impact on resident self-direction as assessed by both residents and staff. For example, the development of a residents' council can have benefits in this area, even though residents emphasize the social and expressive functions of such councils more than their role in facility governance and policy formation (Ryden, 1985). When staff members allow residents to make and implement important decisions, such councils are more likely to contribute to residents' sense of control.

Finally, to maintain an emphasis on independence, staff must be aware of its importance to residents and refrain from interfering with its exercise. Settings with greater staff resources tend to develop more emphasis on resident self-direction, as do facilities that provide staff with their own separate areas. By identifying how particular staff characteristics are related to social climate, it may be possible to develop focused interventions. Thus, in-service training can deal with how staff may subtly undermine resident efforts at independence. Program managers can help staff recognize how to encourage independent functioning and self-care rather than dependency (Stirling & Reid, 1992).

Maintaining Organization

Administrators generally strive to have a well-organized facility in which staff carry out their duties efficiently and in which participants are clear about rules and expectations. Achieving good organization is a greater challenge in large facilities and in those with residents who have fewer social or functional resources. However, administrators can avoid some of the negative effects of large size by intervening in specific processes in the program, for example, by taking steps to reduce the facility's impersonal qualities. In addition, the presence of older, more experienced staff members contributes to organization, so that greater effort might be made to recruit and retain such staff.

Although physical features and policies and services are not strongly related to facility organization, more space for resident functions and more privacy are associated with better organization. Managers may be able to enhance organization by reducing crowding and encouraging respect for residents' privacy needs.

Enhancing the Social Climate

Our suggestions for improving social climate in group residences imply an active planning role for both residents and staff. To implement such changes effectively, we need better staff training and staff with more experience in psychological and behavioral areas. Winnett (1989) has emphasized the important role of mental health practitioners in promoting an environment oriented toward rehabilitation. Psychologists and other similarly trained staff can work with residents and staff to assess and change specific aspects of the social climate to promote optimum resident and staff functioning.

As well as the end product, the process of identifying goals and planning change can benefit residents and staff members, who may often feel relatively powerless within a facility. As we have seen, although residents and staff may have somewhat different perspectives, their interests overlap; the majority of factors that contribute to residents' perceptions of harmony, resident self-direction, and good organization contribute similarly to staff perceptions.

Having shown that aggregate resident and staff characteristics, physical, and policy and service factors influence the social climate of a facility, in the next two chapters we consider how these program characteristics affect residents' adjustment, involvement in activities, and use of facility services.

8

Personal Control Policies, Social Climate, and Residents' Adaptation

In previous chapters, we emphasized the importance of policies that empower residents and of social climates that are harmonious and orderly. This emphasis reflects not only the widely shared cultural value placed on individual autonomy and friendly relationships (Ryff & Essex, 1991), it also reflects a body of research documenting the positive impact of these environmental factors on resident well-being (for reviews, see Carstensen, 1991; Rodin et al., 1985; Schulz, Heckhausen, & Locher, 1991).

In this chapter, we evaluate the effects of facility policies and social climate factors on resident adaptation. Specifically, we ask whether residents in our sample of 262 community facilities show better adaptation when policies empower residents and when social climates emphasize support, resident self-direction, and order.

We also address two related issues here. The first is whether groups of impaired residents can benefit from policies enhancing resident autonomy. Apparently, many administrators and staff believe that they cannot; consequently, nursing home residents have fewer opportunities for empowerment than do residents in other types of facilities. Accordingly, we ask whether residents' aggregate functional abilities moderate the positive relationships described above. For example, in a facility with better-functioning residents, is high choice more closely associated with resident adjustment than it is in a facility with more-impaired residents?

The second, related issue concerns a possible trade-off in benefits to residents versus staff. Willcocks and her colleagues (1987) found that features benefiting residents may have a negative impact on staff. Accord-

This chapter is coauthored by Christine Timko.

ingly, here we ask how facility policies and social climate influence staff turnover and rated functioning.

THE INFLUENCE OF PERSONAL CONTROL POLICIES AND SOCIAL CLIMATE

We propose that facility policies can empower residents in four major ways: (1) by allowing them more choice in their daily routines, (2) by giving them more formal control over the facility's programs and policies, (3) by giving them more privacy, and (4) by developing means to communicate policies to residents. We refer to these four as personal control policies.

To examine relationships between personal control policies and resident adaptation, we conducted regression analyses. In the first step, we entered the level of care and residents' social resources and functional abilities; by controlling for these three factors, we take into account associated differences in resident outcomes before we look at the relationships between personal control policies and resident adaptation. In the second step of the analyses, we entered four indices that tap personal control policies: policy choice, resident control, privacy, and policy clarity.

Our analyses included five indicators of adaptation. First, Resident Adjustment, a five-item Rating Scale dimension, summarizes observers' evaluations of the appearance, activity level, and interaction among residents.

Two indices assess residents' engagement. Activities in the Community is a 14-item RESIF scale that measures residents' participation in activities that take them outside the facility. Participation in Facility-Planned Activities comprises 13 items that cover various kinds of organized activities in the facility. We include both measures of involvement because they are likely to have different determinants and different consequences as well.

Use of Health Services (8 items) and Daily Living Assistance Services (14 items) measure the extent to which services provided by the facility are used by its residents. When level of care and functional ability are controlled, less reliance on facility services is likely to reflect appropriate use of services and better adaptation. Information on involvement in facility-planned activities and services was obtained with the Policy and Program Information Form (POLIF); the staff who completed information on the availability of activities and services also answered questions about average use.

To examine the influence of social climate on resident adaptation, we took a similar approach. First, we controlled for level of care, social resources, and functional abilities. Then we entered the indices that tap the harmony, resident self-direction, and order of the social climate as predictors of resident adaptation. These social climate factors are the Sheltered Care Environment Scale (SCES) subscales of Cohesion, Conflict, Independence, Resident Influence, and Organization (chapters 5 and 7).

Table 8.1 shows the partial correlations between the predictors and each of the five indicators of adjustment. The partial correlations control for level of care, social resources, and functional abilities.

Level of Care, Social Resources, and Functional Abilities

As expected, residents in facilities that provide more care are more likely to rely on facility services and to participate in facility activities than are residents in lower levels of care, even after taking into account differences in their functional abilities and social backgrounds.

Facilities with more socially advantaged residents tend to have slightly better outcomes; residents in these facilities participate in more activities in the community and are viewed by outside observers as better adjusted. In facilities with more functionally able residents, residents are rated as better adjusted and are more involved in activities in and outside the facility (Table 8.1).

Personal Control Policies and Resident Adaptation

After level of care, social resources, and resident functional abilities are controlled, facilities with higher levels of policy choice, resident control, privacy, and policy clarity have better rated resident adjustment and higher levels of resident involvement in community activities. In addition, residents in these facilities rely less on the facility's health and daily living services. Participation in facility activities also is lower in facilities where residents have more choice and control and where policy clarity is higher.

Thus, residents in facilities with policies emphasizing personal control are more involved in the surrounding community and are seen as active and engaged. Although more residents are involved in community activities in these facilities, fewer participate in facility-organized activities. Facilities that take into account individual preferences in scheduling also appear to tailor their services and activity program more to individual needs. They are less likely to provide services and activities to entire groups of residents, and, as a consequence, provision of unneeded services and unwanted activities is minimized.

These findings extend the results of intervention studies (e.g., Slivinske & Fitch, 1987) by showing that stable, high levels of personal control are associated with better adjustment among elderly residents of community care settings (Moos, 1981; Timko & Moos, 1989). Although short-term changes in opportunities for exerting personal control may be highly salient (Schulz & Hanusa, 1980), ongoing levels, as reflected in facility policies, are likely to have important long-term effects on elderly residents' adaptation.

Table 8.1 Partial and multiple correlations from regression analyses predicting resident adaptation from level of care, suprapersonal factors, personal control policies, and social climate ($N = 262$ facilities)

	Resident adaptation				
	Resident adjustment	Activities in the community	Facility-planned activities	Health services	Daily living assistance services
LEVEL OF CARE	−.08	−.07	.29***	.35***	.36***
SUPRAPERSONAL FACTORS					
Social Resources	.17**	.13*	.08	−.05	.10
Functional Ability	.19**	.30***	.20**	.04	−.02
FACILITY POLICIES					
Choice	.17**	.21***	−.19**	−.23***	−.30***
Control	.25***	.20**	−.16**	−.22***	−.22***
Privacy	.29***	.24***	−.11	−.36***	−.29***
Clarity	.29***	.21***	−.16**	−.23***	−.19**
SOCIAL CLIMATE					
Cohesion	.46***	.12	−.02	−.17**	−.08
Conflict	−.02	−.03	−.12*	.02	.02
Independence	.39***	.23***	−.23***	−.30***	−.26***
Resident Influence	.36***	.11	−.07	−.26***	−.05
Organization	.31***	.10	.17**	.00	−.01
OVERALL R	*.75****	*.71****	*.46****	*.71****	*.71****

Partial correlations control for level of care, social resources, and functional ability; overall R includes level of care.
$^{*}p < .05$; $^{**}p < .01$; $^{***}p < .001$.

Social Climate and Resident Adaptation

Even after considering residents' social resources and functional abilities, the social climate dimensions are predictably associated with indices of adaptation (Table 8.1).

Cohesion and Conflict

In facilities that residents rate as high on cohesion, outside observers rate resident adjustment higher, residents are more likely to be involved in activities in the community, and use of facility health services is lower. Where more conflict is reported, fewer residents participate in facility activities.

Individuals generally benefit from being involved in warm, supportive relationships (Antonucci, 1990; Carstensen, 1991). The present findings point to the importance of such relationships within congregate residential settings. Open expression of anger appears to have less impact than does warmth and supportiveness, except that residents may avoid organized social activities if they perceive a high level of conflict in relationships.

Independence and Resident Influence

The results for perceptions of resident self-direction parallel those for personal control policies. In facilities that residents see as emphasizing resident independence, residents are rated as having better adjustment and are more likely to be involved in activities in the community. They also participate less in facility activities and rely less on facility services. In facilities with a high level of resident influence, residents are rated as better adjusted and are less likely to use the available health services.

As we have seen, residents in facilities with policies that offer more personal control tend to report having more independence and influence (chapter 7). Thus, policies that promote personal control may influence residents in part because they help to create a resident-directed social climate.

Organization

Facilities that are better organized are rated by outside observers as having better-adjusted residents. In addition, these facilities tend to have a higher level of involvement in planned activities in the facility.

Predicting Resident Outcomes

Previous research has shown that the degree of constraint imposed by an environment influences an individual's beliefs about personal control (Arling, Harkins, & Capitman, 1986; Hulicka, Morganti, & Cataldo, 1975). Older people who believe that they have more personal control tend to report higher activity levels, better health, and more positive self-concept

(Rodin et al., 1985). Thus, although there are exceptions (Burger, 1989), more opportunities for personal control are associated with better physical and psychological well-being (for reviews, see Rodin, 1986; Rodin et al., 1985; Schulz et al., 1991). Our findings are consistent with this body of research.

As expected, outside observers rate as more active and involved those residents with more social resources and intact functioning. Facility policies and social climate are also related to such ratings; where policies provide residents more personal control, where residents have closer relationships and higher levels of self-direction, and where the facility is better organized, outside observers have more positive impressions of resident adjustment.

Involvement in the community tends to be greater among resident groups with higher functioning and more social resources. In addition, in facilities where policies emphasize resident autonomy and the social climate is oriented toward independence, residents are more likely to be involved in activities in the community. These factors may influence one another: Staff may be less inclined to impose restrictions on residents who maintain active involvement in the community, and residents may be more likely to stay involved if facility policies treat them as autonomous individuals.

Involvement in formal activities organized within the facility shows a somewhat different pattern than does participation in community activities. Participation in facility activities tends to be lower where residents exert more choice and control, suggesting that in highly structured, staff-controlled facilities, some residents may be passive or unwilling participants in the activity program. Thus, one explanation for a low level of participation in facility activities is that the activity program is being tailored to individual interests and that residents are exercising appropriate control over their participation. However, we also found that participation is lower when the facility has more conflict and when it is poorly organized. A moderate level of participation may reflect a balance between offering an appealing program of activities and allowing residents choice about participating.

Finally, we found that reliance on facility services is lower in facilities with policies supporting resident autonomy and a social climate oriented toward independence. Again, these factors may be mutually reinforcing: When residents use fewer of the available services, they may expect more choice and control and develop a social climate more oriented toward independence.

Overall, these sets of predictors account for 21% to 56% of the variance in these indices of resident outcomes. Facility policies and social climate added between 7% and 25% of variance (mean = 13%) to that accounted for by level of care and the aggregate resident factors. These findings imply that changes in policies and social climate may have substantial effects on these aspects of residents' outcomes.

With some exceptions, the associations between social climate and the indices of resident adaptation continue to hold after controlling for the four personal control policies in addition to level of care and the aggregate resident factors. Thus, social climate appears to influence resident adaptation independently of facility policies and resident characteristics.

RESIDENTS' AGGREGATE FUNCTIONAL ABILITIES AS A MODERATING FACTOR

Our findings are consistent with expectations that personal control policies and a positive social climate contribute to better adaptation. We now examine the idea that the aggregate functional ability of facility residents moderates the relationship between these facility features and residents' adaptation.

Models of Person–Environment Congruence

Some models of person–environment congruence suggest that groups of high-functioning residents should be able to achieve good adaptation in a broad range of settings and that they have greater adaptive capacity to compensate for adverse environmental conditions than do individuals with impaired functioning (e.g., Lawton, 1989). Thus, high-functioning residents should adapt about as well in facilities with policies that provide low levels of personal control as in those that provide high levels. In contrast, low-functioning resident groups may be more vulnerable to the impact of their immediate environment. They would be expected to adapt poorly in facilities that provide too few or too many opportunities for personal control. High environmental demands, as indexed by policies that enhance control and independence, may have negative consequences for very impaired older people by overwhelming their fragile resources.

Noting that some research points to greater environmental responsiveness of functionally competent individuals, Lawton (1989) expanded his model to focus on personal and environmental resources. Viewed in these terms, policies that enhance personal control and a cohesive and resident-directed social climate would have the greatest benefit for groups of functionally able residents. In contrast, functional capacity is the primary determinant of impaired residents' adjustment and activity levels. Thus, as residents' functional abilities decline, variations in policies and resident self-direction should be less important. Overall, personal control and independence should have more positive consequences for groups of older people who are functioning well than for those who are functioning poorly.

This view is more fully developed in needs-hierarchy models (Carp & Carp, 1984). In this model, higher level needs (such as the need for control) are activated only when lower level needs (such as life maintenance) are satisfied. Thus, well-functioning residents, whose lower level needs are likely to be satisfied, should show better adaptation in facilities that give them more opportunities for control and poorer adaptation in facilities in which they have fewer such opportunities. In facilities with severely impaired populations, however, the presence of a residents' council or other opportunities for control would make little difference for residents' level of engagement or adjustment. Impaired residents presumably would be more concerned about their life-maintenance needs and less influenced by personal control policies.

Although personal control policies may thus have a greater impact on functionally competent residents, other facility attributes, such as the level of interpersonal support, may have greater relevance than do policies to the needs of impaired residents. Within this model, no generalization can be made about the greater vulnerability of any group of residents to the influence of environmental conditions. The match between the individual's needs and abilities and the resources and challenges provided by the environment determines outcomes.

The present data allow us to examine these models and the predictions they generate.

Aggregate Functional Abilities as a Moderator of Personal Control Policies

To identify the moderating influence of functional abilities on the relationship between personal control policies and adaptation and to test the models of person–environment congruence, we conducted multiple regression analyses in which we entered functional abilities and one of the four facility policies shown in Table 8.1, followed by the interaction between functional ability and the facility policy (for the statistical procedures, see Finney et al., 1984). The four facility policies and five dependent variables generated 20 regressions; in 10 of these, the interaction term was statistically significant ($p < .05$).

Rated adjustment and involvement in community activities tended to show similar benefits from policies that support autonomy, regardless of the functional abilities of facility residents. The interaction terms were not significant for these dependent variables, except that functional ability interacted with policy choice to predict resident adjustment.

Where interactions were present, the consistent pattern of findings shows that better-functioning resident groups are more strongly influenced by facility policies emphasizing resident self-direction. Policies that give more control to residents are associated with better adjustment, less involvement in organized activities, and less use of services in settings with functionally intact residents but not in facilities with functionally impaired resident groups. For example, the interactions of functional abilities with policy choice, resident control, and privacy were significant for participation in facility activities and use of health and daily living services.

Functional Abilities, Policy Choice, and Resident Adjustment

In Figures 8.1 through 8.4, *high functioning* refers to facilities that score one standard deviation above the mean on Functional Abilities; on average, their residents can carry out almost all (88%) daily activities without assistance. Most apartment facilities and some residential care settings are at or above this level. *Moderate functioning* refers to facilities at the mean; on average, their residents can carry out over half (61%) of their daily activities without assistance. Most residential care facilities and a few nursing homes score at or above this level. *Low functioning* refers to facilities scoring one standard deviation below the mean; their residents can carry out only about one-third (34%) of their daily activities without assistance. About 40% of the nursing homes score below this level; the remainder fall into the range between low and moderate functioning.

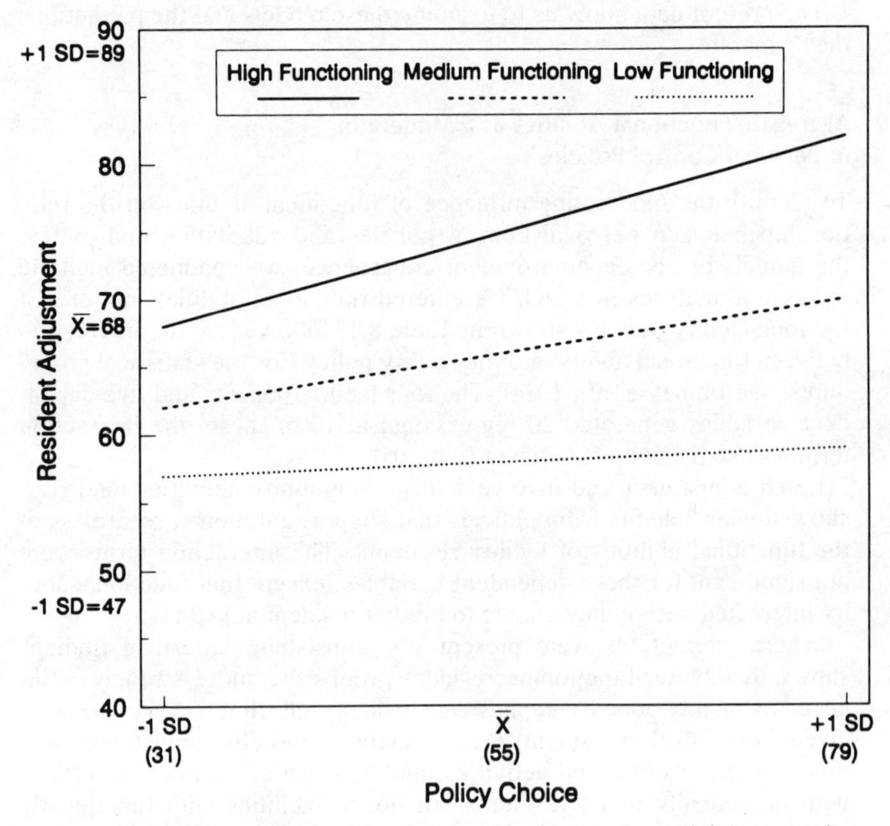

Figure 8.1 Interaction of policy choice and aggregate functional abilities to predict resident adjustment.

As previously noted (Table 8.1), policy choice and aggregate functional abilities are each associated with better resident adjustment. Compared with groups of medium- or low-functioning residents, groups of high-functioning residents are rated as better adjusted at each level of policy choice, as shown in Figure 8.1. Similarly, on average, resident adjustment is rated higher in facilities with more policy choice.

In addition, Figure 8.1 shows an interaction between policy choice and residents' aggregate functional abilities. Resident adjustment tends to be higher in facilities that provide more choice and in which resident functioning is moderate or high. In facilities with more impaired residents, however, resident adjustment is independent of the level of choice. High-functioning residents in low-choice settings (one standard deviation below the mean, equivalent to 31% on a 100-point scale) obtain an average score of 68% on adjustment. In high-choice settings (one standard deviation above the mean or 79%), they obtain an average score of 82% on adjustment. In contrast, poorly functioning residents score about 58% on adjustment regardless of the level of policy choice.

Functional Abilities, Policy Choice, and Daily Living Assistance

The interaction shown in Figure 8.2 similarly shows higher-functioning resident groups to vary more in outcomes in relation to the level of choice. In facilities with intact or moderately impaired residents, more policy choice is related to less reliance on daily living services. In facilities with very impaired residents, policy choice has almost no effect on the use of services. More specifically, high-functioning residents use about 69% of the facility's daily living assistance in low-choice settings and only 41% in high-choice settings. In contrast, very impaired residents use about 69% of the facility's daily living assistance irrespective of the level of policy choice.

The same basic pattern applies to the impact of policy choice on use of health services and of resident control and privacy on use of health services and daily living assistance. In general, higher levels of personal control policies are related to less use of facility services for relatively intact and moderately impaired resident groups, but the levels of these policies have little or no influence in facilities with very impaired residents. Facilities whose policies constrain resident autonomy may provide services to higher-functioning residents who do not need those services. In contrast, in facilities serving impaired resident groups, service use is not related to policies.

Aggregate Functional Abilities as a Moderator of Social Climate

To examine the moderating influence of functional abilities on the relationship of facility social climate to adaptation, we conducted similar multiple regression analyses in which we entered functional abilities and one of the five social climate dimensions shown in Table 8.1, followed by the

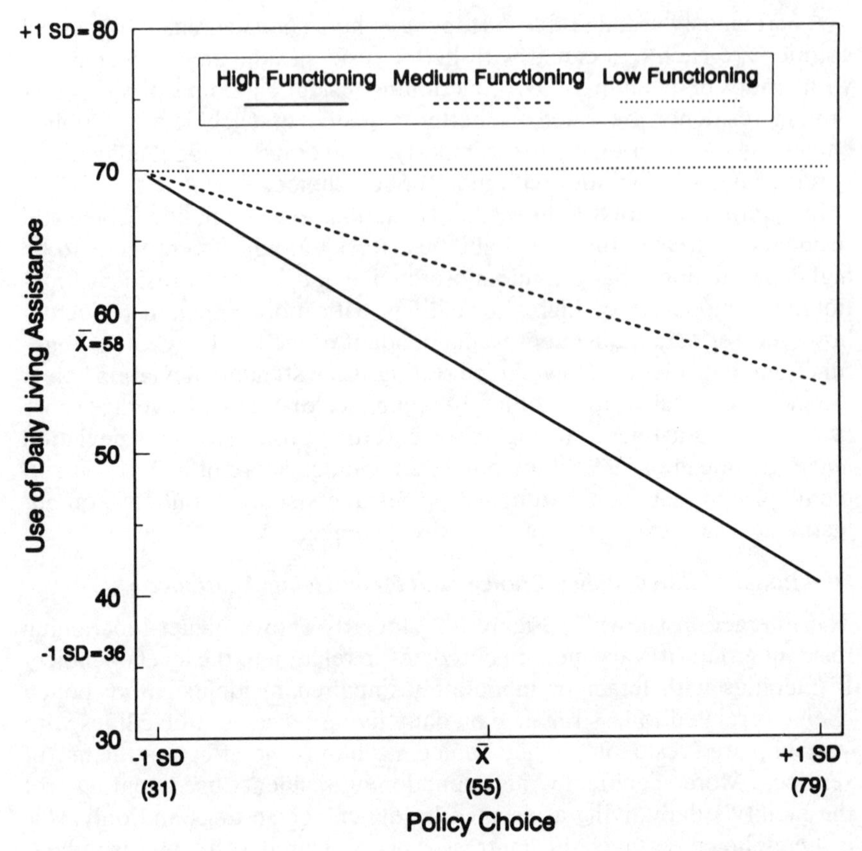

Figure 8.2 Interaction of policy choice and aggregate functional abilities to predict use of daily living assistance.

interaction between functional abilities and the social climate dimension. The five social climate dimensions and five dependent variables generated 25 regressions; in six of these the interaction term was significant ($p < .05$). Independence shows a significant interaction with functional abilities for all five of the outcome criteria.

In facilities with high-functioning residents, resident adjustment and involvement in activities in the community tend to be higher, participation in facility-planned activities lower, and use of health services and daily living assistance lower when independence is emphasized. The emphasis on independence is also associated with better adjustment in settings with impaired residents; however, it is not related to participation in facility-planned activities, involvement in activities in the community, or service use among impaired resident groups. Figures 8.3 and 8.4 illustrate some of these results.

In addition, compared with groups of impaired resident, the adjustment

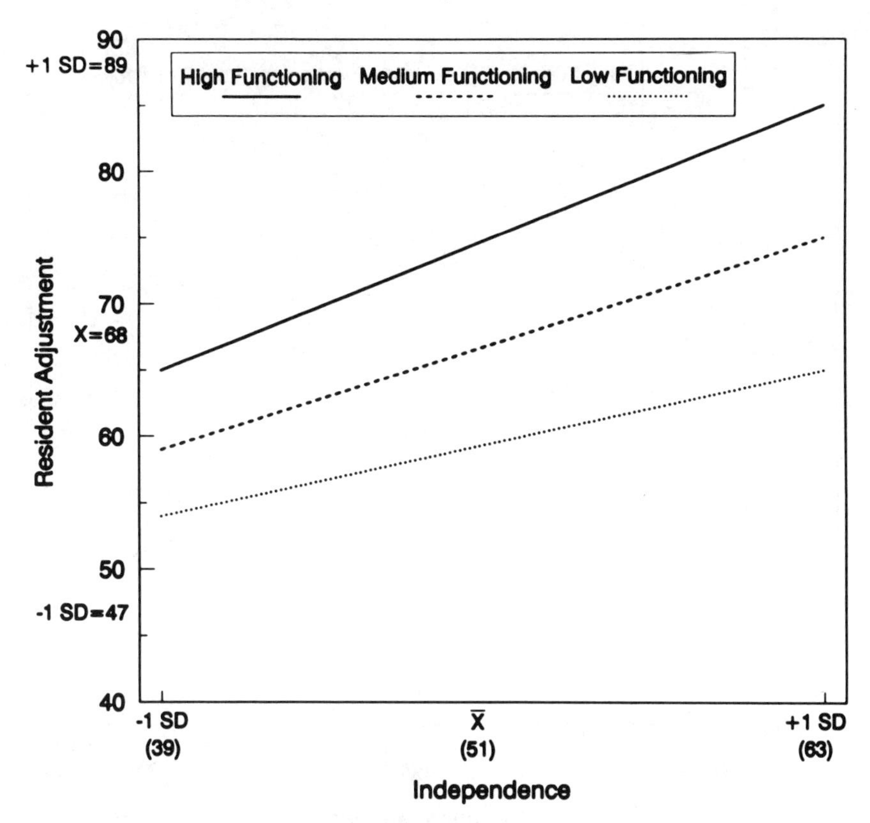

Figure 8.3 Interaction of independence and aggregate functional abilities to predict resident adjustment.

of high-functioning resident groups is also more strongly related to the facility's level of conflict. Conflict appears to have relatively little impact on adjustment among moderate- or low-functioning resident groups. In contrast, high-functioning resident groups with high levels of conflict are seen as less well adjusted.

Functional Abilities, Independence, and Resident Adaptation

As shown in Figure 8.3, resident adjustment increases as independence increases for all facilities, but the rate of increase is highest in facilities with well-functioning residents and least in facilities with functionally impaired residents.

Functional Abilities, Independence, and Use of Services

Figure 8.4 shows the interaction of functional abilities and independence in predicting use of daily living assistance services. As independence increases, use of daily living services declines in facilities with high- and medium-functioning residents. In contrast, the level of independence is

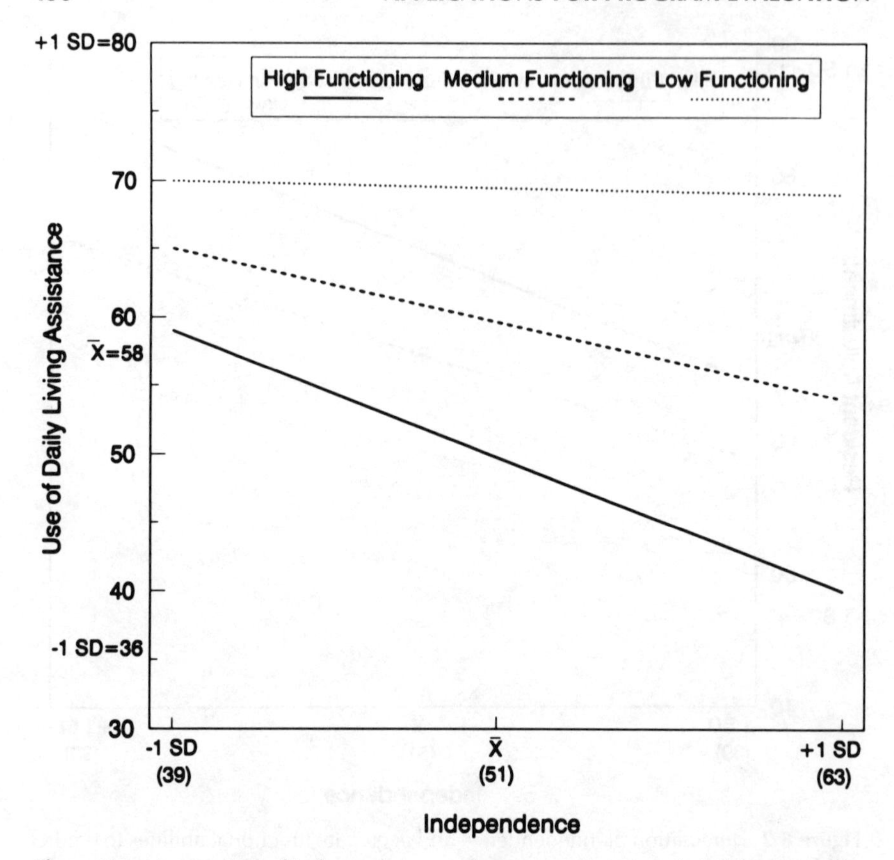

Figure 8.4 Interaction of independence and aggregate functional abilities to predict use of daily living assistance services.

not related to reliance on daily living services in facilities with impaired residents.

The same pattern emerged for participation in facility-planned activities and for use of health services. Thus, an increased emphasis on independence may reduce reliance on facility services in facilities with residents who are functioning relatively well, whereas it has little impact in facilities with residents who are more impaired.

The findings for social climate are quite consistent: An orientation toward independence is strongly associated with better adjustment and less use of services in facilities with relatively intact residents. Emphasis on independence also appears to benefit adjustment among impaired resident groups but does not affect their involvement in facility-planned activities or use of health and daily living services. In general, these findings support the idea that functionally competent residents are better able to take advantage of environmental opportunities for control.

Implications for Models of Person–Environment Congruence

These results are consistent with a needs-hierarchy model in which the need for autonomy and control is most strongly activated for high-functioning residents, who are most able to meet their own life-maintenance needs. Thus, high-functioning resident groups show better adaptation when policies support personal control and the social climate emphasizes independence. These policy and social climate factors show some benefits for low-functioning resident groups but are generally less salient for them.

In contrast, supportive aspects of the setting, such as a cohesive social climate and good organization, seem to have benefits regardless of the group's level of functioning. The needs for warm relationships and order are activated across the broad range of functioning found in these congregate residential settings.

THE IMPACT OF POLICIES AND SOCIAL CLIMATE ON STAFF PERFORMANCE

Thus far, we have focused on resident outcomes; in this section, we briefly consider the impact of facility policies and social climate on staff performance. Group residential facilities are labor-intensive settings that require the efforts of numerous staff members. In particular, nursing homes have almost as many staff members as residents. Strahan (1987) estimated that, in U.S. nursing homes alone, more than 1.1 million full-time employees are involved in providing direct and indirect services. Many of these employees experience their work as stressful and demeaning. For example, staff in nursing homes often engage in menial and repetitive tasks, are supervised closely, and receive few social or economic rewards (Tellis-Nayak & Tellis-Nayak, 1989). Such difficult work conditions may contribute to low staff morale, impersonal care, and high turnover.

Staff turnover is affected by such factors as salaries and benefits, workload, and the current labor market. We believe that staff turnover is also related to the facility's personal control policies and social climate (Moos & Schaefer, 1987). Conditions that contribute to making the setting a positive environment for residents, such as harmony, self-direction, and good organization, should also improve staff morale and consequently reduce staff turnover. However, some theorists have argued that there may be an inherent conflict of interest between the needs of residents and staff, which results from the way that congregate care is presently organized. For example, Willcocks and her colleagues (1987) found that in homes where the policies were more oriented toward resident needs, staff tended to worry more.

We examined two measures of staff performance—staff turnover and outside observers' ratings of staff functioning—and their correlates in the 262 community facilities. Staff turnover is the proportion of staff who have been employed in the facility for less than a year. Staff Functioning, a five-item Rating Scale dimension, summarizes observers' evaluations of the availability of staff and the quality of their interactions with residents and one another. As we did when examining resident outcomes, we first controlled for level of care, residents' social resources, and residents' functional abilities; then we examined the relationships between facility factors and staff outcomes. Concerning level of care, we found that nursing homes score highest and apartment facilities lowest on ratings of staff functioning, whereas level of care is not related to staff turnover once residents' aggregate characteristics are considered. In facilities in which residents have more social resources, staff turnover is lower and ratings of staff functioning are higher. Resident functional abilities are not related to these staff outcomes.

Personal Control Policies and Staff Performance

Personal control policies are not associated with either ratings of staff functioning or the turnover rate for staff. Thus, facility policies aimed at recognizing residents' needs for autonomy, which tend to benefit resident functioning, appear to have little impact on the work conditions of staff. In particular, these results suggest that residents can gain these benefits without increasing the level of strain on staff or compromising their level of organization and efficiency.

Social Climate and Staff Performance

Several of the social climate dimensions are related to ratings of staff functioning and to staff turnover, independently of level of care, residents' social resources, and residents' functional abilities.

Cohesion and Conflict

Ratings of staff functioning are higher in settings in which staff and residents report higher cohesion. Staff turnover is higher where staff report more conflict. Interpersonal conflict and lack of group cohesion appear to interfere with the quality of staff performance and their motivation to continue working in a facility. These results are consistent with the findings in nursing homes and acute care settings, where these social climate characteristics are related to poor job morale and lack of satisfaction (e.g., Constable & Russell, 1986; Grau et al., 1991; Revicki & May, 1989).

Independence and Resident Influence

Staff functioning is rated higher in settings where residents see themselves as having more independence and more influence. Because our data show that staff functioning and turnover are unrelated to residents' functional abilities, they support the view that high levels of resident impairment and the consequent physical care routines do not inevitably lead to demoralization and high staff turnover. Rather, programs that do not value residents' input and independence may be costly to staff performance.

Resident influence, per se, may not have a direct impact on staff functioning but may reflect the management emphasis on centralized, autocratic versus decentralized, democratic decision making. Consistent with this pattern, Waxman, Carner, and Berkenstock (1984) found more turnover among nursing assistants in nursing homes that limited staff's autonomy and decision-making power. The researchers concluded that turnover may be prompted by more rigid, centralized policy making. Similarly, Stirling and Reid (1992) found that a staff training program designed to encourage participatory control between staff and residents resulted in improved resident self-concept, in their viewing the environment as more supportive and less controlling, and in increased staff perceptions of control.

Organization

We found poorer staff functioning and higher turnover in poorly organized facilities. These factors are likely to be mutually reinforcing. Staff are more satisfied when they work in a clear and predictable setting, and lower staff turnover may contribute to efficiency and predictability.

The influence of organization may vary in line with associated levels of cohesion and self-direction. Without cohesion and self-direction, an emphasis on organization may be experienced as restrictive and controlling. In the context of harmonious relationships among staff and a climate of independence, however, organization tends to be appraised as enhancing clarity and facilitating shared goals and activities. Thus, staff functioning may be higher and staff turnover lower in facilities that balance order and predictability with staff and resident input. We obtained just such a finding in the veterans nursing units (Brennan & Moos, 1990).

These findings imply that the social climate can have consequences for staff as well as for residents. Where relationships are warm and supportive, outside observers report better staff functioning; staff also are more likely to stay in their job in facilities in which staff report low levels of conflict. The quality of facility organization reported by staff and residents appears particularly important in contributing to staff job performance and commitment. In addition, a decentralized or participative organization in which management seeks input from residents and staff seems to improve staff functioning and to facilitate individualized patterns of care. Overall, resident and staff interests do not seem to be at odds. In particular, a social climate emphasizing resident self-direction does not appear to be detrimental to staff.

IMPLICATIONS FOR INTERVENTION AND EVALUATION

Policies that promote residents' choice, control, and privacy and a cohesive, independence-oriented, well-organized social climate are related to better resident adjustment and to greater resident involvement in community activities. In addition, policies and a social climate that emphasize personal control are associated with less reliance on facility services. Staff seem to benefit from a social climate that emphasizes warm relationships, good organization, and resident involvement; these benefits are reflected in ratings of staff functioning and, to a lesser extent, in staff turnover rate.

Furthermore, personal control policies and social climate tend to influence some types of resident groups more strongly than others. In facilities with relatively intact residents, personal control policies and an independence-oriented social climate are strongly associated with better adjustment and less use of facility services. In facilities with residents who have significant impairment, however, these policy and social climate factors have little or no influence on ratings of resident adjustment, participation in facility-planned activities, and use of facility services.

Policies that promote personal control may be especially important for residents in residential care facilities, who typically are able to function somewhat independently in their daily activities but who are given relatively little policy choice and who show high levels of reliance on daily living assistance and health services. For example, in Maude's Community Care Home residents perform about 75% of their daily activities without assistance, but they are given choice in only about 25% of the areas included in the Policy Choice scale (for the POLIF profile for Maude's, see chapter 4). About half of the residents use the health services available in this facility and almost all use the daily living services that are offered. Our findings suggest that an increase of about one standard deviation in policy choice (up to about 50%) might be associated with a decline of about 10% in residents' reliance on daily living assistance and health services.

An active and engaged resident population can be supported by developing policies that recognize residents' need for control and by creating a social climate with an emphasis on warm relationships, resident independence, and good organization. Some of these policy and social climate factors have less influence on impaired resident groups.

Participation in facility-planned activities may be an indication of better adaptation for very impaired resident groups, who are less able to go into the community for activities there. Good organization and provision for privacy facilitate their involvement. In contrast, for groups of residents who are more intact, involvement in facility activities is less when residents have more opportunities for choice and control and when there is more emphasis on independence; intact resident groups in these facilities tend to participate more in activities in the community. Thus, in facilities serving impaired residents, involvement in facility activities may be viewed as a positive outcome, but in facilities serving less impaired populations, such involvement may be attained only at the cost of reductions in self-initiated activity and quality of life.

When planning facility-level interventions for impaired residents, staff need to balance the level of emphasis on support and independence. Baltes and her colleagues (Baltes, 1988; Baltes & Reisenzein, 1986; Baltes & Wahl, 1992) have identified a "dependency/support script" whereby staff may inappropriately reward residents' dependent needs. Some impaired residents respond by manifesting overly dependent and helpless behavior to solicit attention and assistance from staff; such behavior can help residents maintain a semblance of personal control over the environment. To counteract these processes, staff need to learn how to provide support to impaired residents in a way that enables the residents to maintain their independent behavior.

More generally, programs may provide different combinations of stressors and resources that allow staff to function effectively. For example, Lyman (1990) used the Physical and Architectural Features Checklist (PAF) and the Policy and Program Information Form (POLIF) as guidelines to help identify the major sources of work-related stressors in two day-care programs for cognitively impaired patients. In one program, the main source of work stressors was the limited physical facilities, such as lack of space and unsafe conditions. However, the program's high structure and support enabled staff to manage the patient care demands. In contrast, in the other program, work stressors stemmed mainly from patient-oriented policies that limited staff control over their work conditions. But because these policies also fostered high expectations for patient functioning and normalized patient–staff relationships, patients became an important source of social support for staff. Thus, enhanced autonomy for patients also benefited staff.

Enhancing Residents' Personal Control and Support

A number of researchers have successfully increased short-term levels of personal control in group residential settings (e.g., Langer & Rodin, 1976; Schulz, 1976; Slivinske & Fitch, 1987). According to Teitelman and Priddy (1988), these studies point to some generic methods that can promote personal control and a harmonious social climate (see also Clark, 1988). One main approach is to reduce stereotypes that engender helplessness and to convince residents that they share responsibility for maintaining their own health and morale. A related approach is to increase residents' ongoing levels of choice and predictability by allowing them some latitude in planning their daily activities and by involving them in some meaningful functions in the facility. Thus, in high-choice nursing homes, residents' satisfaction is comparable to that for older residents of the community, whereas residents in low-choice nursing homes report lower life satisfaction (Vallerand, O'Connor, & Blais, 1989).

Residents who are given more responsibility may alter their interactions with staff and other residents, and this may trigger a set of mutually beneficial changes. For example, when Banziger and Roush (1983) gave nursing home residents more responsibility and a chance to care for bird feeders, the residents were happier and more active. Similarly, Winkler and her colleagues (1989) found that nursing home residents interacted more with each other and with staff after introduction of a live-in dog but that residents' behavior reverted to baseline after a few weeks, perhaps in part because staff assumed control over the care and feeding of the dog.

Many control-enhancing innovations have only short-term effects (Okun, Olding, & Cohn, 1990). Long-term benefits for residents are likely to depend on interventions that contribute to stable increases in personal control policies and support (Rodin, 1986). For example, Agbayewa and colleagues (1990) tried to empower long-term care residents by instituting a resident–staff group to participate in unit management. The regular weekly meetings were cochaired by an elected resident and staff member. The agendas included issues that residents or staff raised, and participants tried to discuss all issues and develop plans to implement desired changes. The group could designate a subcommittee to develop and monitor implementation plans and report back to the general meeting. These open community meetings led to improved resident–staff relationships, increased resident autonomy, less resident problem behavior, less need for psychiatric consultation, and better staff morale. Resident councils may also provide residents with a sense of control and lead to some improvements in residents' well-being and quality of care in a facility (Coppola et al., 1990; Meyer, 1991).

Managers and staff can adapt a generic four-step intervention plan to implement some of these changes (Teitelman & Priddy, 1988). First, residents make a list of potentially alterable problems in their facility and rate how much these problems bother them. Second, residents describe what they would do to manage these problems. Third, managers, staff, and residents intervene by making appropriate changes in the facility. When problems persist despite efforts to modify them, residents can try to learn more effective coping strategies. Fourth, follow-up evaluation is conducted to document the changes that have taken place and to plan additional interventions.

We focus more on planning intervention programs in chapter 12. Next, we examine the influence of personal control policies and social climate on individual residents.

9

Residents' Engagement in Activities

Active use of leisure time is commonly viewed as a sign of healthy adaptation at all stages of the life cycle. Activity involvement can provide variety and stimulation, social roles that preserve positive self-concept, and interpersonal contacts that provide support and maintain social skills. Empirical evidence generally confirms the positive nature of active involvement (Larson, 1978). For example, a social environment that facilitated resident activity seemed to promote longevity among residents of a congregate apartment facility (Carp, 1978–79). High activity levels also predicted longevity among residents of nursing homes (Stones, Dornan, & Kozma, 1989).

In congregate residential settings, assumptions about the benefits of continued engagement lead to an emphasis on organizing and facilitating activities for residents. In chapter 8, we examined the influence of facility features on aggregate resident outcomes, including their involvement in community activities and participation in facility-planned activities. In this chapter, we focus on personal characteristics and facility features that are related to individual residents' involvement in activities.

DETERMINANTS OF ACTIVITY INVOLVEMENT

Figure 9.1 presents a model of the program and personal determinants of individual residents' activity involvement. This model is a variation of the conceptual framework introduced in Figure 1.1. In brief, the model in Figure 9.1 shows that objective program characteristics, such as aggregate resident and staff characteristics and personal control policies (panel

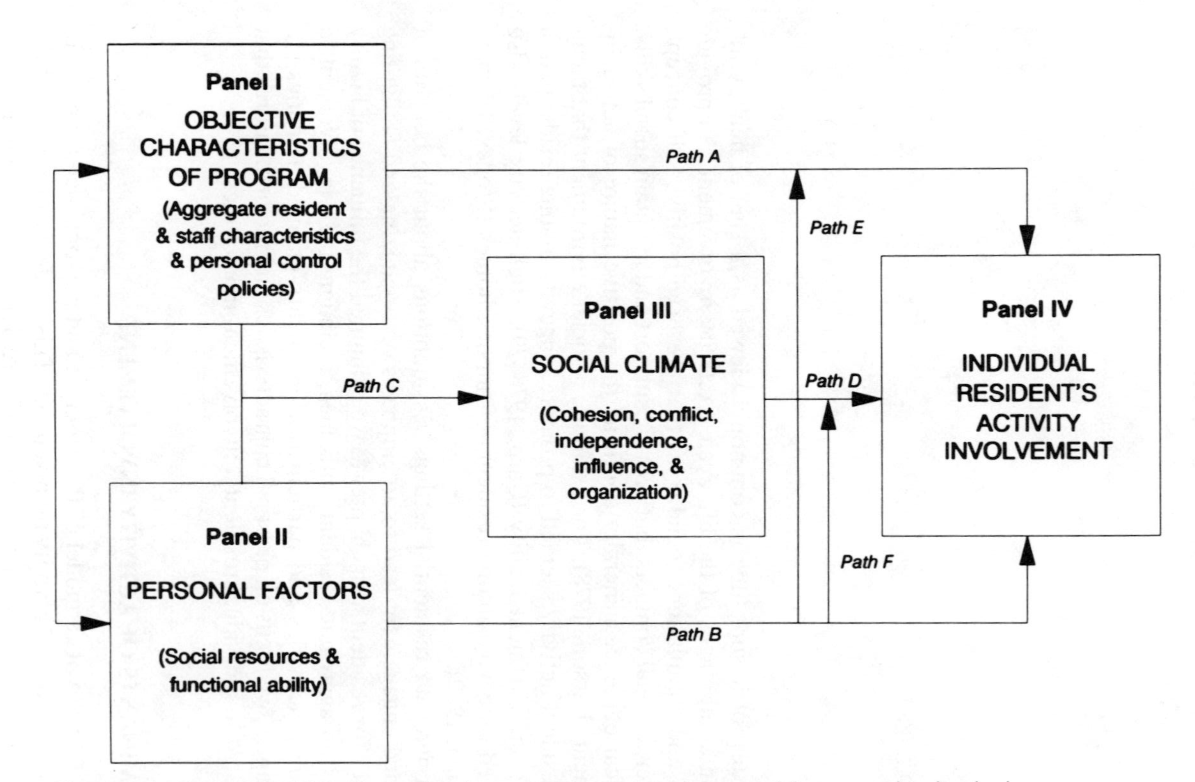

Figure 9.1 A model of the relationship between program and personal factors and individual resident's activity involvement.

I), can influence individual residents' activity involvement (path A to panel IV). For example, activity involvement tends to increase following interventions to enhance residents' personal control (Rodin et al., 1985). Similarly, social and ethnic homogeneity appears to improve resident social interaction (Bergman & Cibulski, 1981; Hinrichsen, 1985).

Individuals also develop different patterns of leisure activity related to their capacities, interests, and prior experience (path B linking panel II and panel IV). For instance, older people who are in better health tend to be more active (George, 1978; Lawton, Moss, & Fulcomer, 1986–87), and nursing home residents who are mobile and mentally intact are more active and show greater behavioral diversity (Brent, Brent, & Mauksch, 1984; Lemke & Moos, 1984).

These program and personal factors can influence the social climate (path C to panel III). In terms of program factors, policies that give residents personal control contribute to residents' sense of self-direction (see chapter 7), and active groups of residents tend to enhance social cohesion, provide role models that promote participation, and invite and motivate other residents to engage in activities. In terms of personal factors, a resident with more social and functional resources may contribute to the development of a cohesive social climate. In turn, a social climate that is harmonious, resident-directed, and well-organized (panel III) may influence individual residents' activity involvement (path D).

Finally, a resident's functional ability may moderate the relationship between objective characteristics of the program and individual outcomes (path E) or between social climate and individual outcomes (path F). As noted in chapter 8, several models of person–environment interaction have been developed; they vary in their predictions about how individuals who differ in their functional abilities may respond to a given environment.

We use the model shown in Figure 9.1 to focus on three main questions that extend to the individual level the facility-level analyses presented in chapter 8: (1) How are the individual's personal characteristics, such as gender, educational background, and functional ability, related to activity involvement? (2) Is activity involvement associated with the aggregate characteristics of facility residents, personal control policies, and social climate factors? (3) How do specific combinations of individual functional ability and program characteristics influence activity involvement?

As we saw in the chapter 8 facility-level analyses, determinants of activity involvement are different for community activities and facility-planned activities; for example, facilities with an emphasis on personal control policies show more resident involvement in community activities and less participation in organized facility activities. In our individual-level analyses, we include measures of involvement in three kinds of activities: informal activities in the facility, activities in the community, and formal activities in the facility.

INDIVIDUAL RESIDENTS' ACTIVITY LEVELS

The Sample of Residents and Facilities

In this part of our research, we use data from individual residents in some of the facilities we studied using the MEAP. We asked 1,428 residents in 42 facilities to complete the Background Information Form (BIF), which asks about their demographic characteristics, current functioning, and involvement in informal activities in the facility and community. Because we wanted to use these data to examine how personal and program factors influence individual residents' activity levels, we studied diverse facilities, each with a varied sample of participants. The overall sample includes 155 residents from 11 nursing homes, 546 residents from 13 residential care facilities, and 727 residents from 18 congregate apartments. A subset of 729 residents in 24 of these facilities also provided information about their participation in facility-organized activities.

The average age of the 1,428 residents was 76; 69% were women, and 77% had lived in the facility for more than a year. Participants in the three types of facilities were of about the same average age, but participants in the apartments were more likely to be women (80%) and to have lived in the facility for more than a year (80%). These respondents reflect the variety of residents in these facilities, although the requirements of completing the questionnaires eliminated very impaired individuals and those with low intellectual functioning.

On average, these 42 facilities differ from the 262 facilities in the larger normative sample of community facilities. More than 40% of the facilities in this smaller sample are congregate apartments and only 26% are nursing homes, compared with 26% apartments and 52% nursing homes in the normative sample. As a consequence, residents in the 42 facilities are, on average, more functionally able and the facilities are higher than the normative sample on personal control policies, such as provision for privacy, policy choice, and resident control. In addition, the 42 facilities are somewhat higher on independence and organization, and, on average, their residents are more involved in community activities and in informal activities in the facility but are less likely to participate in facility-organized activities.

Sociodemographic Factors and Functional Ability

Each respondent completed the BIF, which contains questions covering demographic characteristics (such as age, sex, and education), ability to carry out daily activities, and participation in informal activities in the facility, activities in the community, and facility-organized activities (for details, see Lemke & Moos, 1989b). The items and their scoring are generally similar to the corresponding subscales from the Resident and Staff Information Form (RESIF), which tap facility-level data (chapter 2).

Focusing on the individual's level of functioning, the BIF Functional Ability subscale has 12 items. Scores are computed as the percentage of items on which the individual indicates independence in daily activities and freedom from physical disabilities. The items include nine daily activities scored for responses of "no help needed." Additional items cover the presence of visual and hearing problems and the need for medications. The majority of the 1,428 residents were free from these disabilities and could perform most of these daily living activities independently (mean = 83%), but there was substantial variability in the sample (*SD* = 17.5%).

Activity Involvement

The BIF includes three activity dimensions, which categorize activities by their formality and location: (1) Activity Level measures involvement in informal activities that take place in the facility; (2) Activities in the Community focuses on activities outside the facility; and (3) Facility-Planned Activities measures participation in activities planned by and taking place in the facility.

The BIF Activity Level subscale measures the participation rate for 13 informal activities in the facility, such as watching TV, playing cards, and doing a hobby; it parallels the RESIF subscale with the same name. The score reflects the percentage of these activities in which the resident was involved during the preceding week.

On average, the 1,416 residents who completed this information indicated that they had participated in more than half of these informal activities (mean = 58%; *SD* = 17%). For example, most had watched television, read a newspaper or book, visited with other residents, and taken a walk. Fewer sewed or knitted, played cards or other games, or were involved in a hobby.

The BIF Activities in the Community subscale measures participation in 14 activities outside the facility, such as visiting friends or relatives, eating in a restaurant, or attending church. Like scores on the RESIF counterpart subscale, scores on this BIF subscale reflect the percentage of these activities in which the individual participated at least a few times a year.

Residents left the facility for about half the indicated community activities (mean = 52%; *SD* = 23%). Again, most residents participated in some of the activities, such as visiting friends or relatives, shopping, or eating in a restaurant, whereas fewer went out to attend religious services or for a movie, went to a senior center, or attended a sports event.

The BIF Facility-Planned Activities subscale measures the resident's participation in 13 facility-organized activities such as religious services, exercise groups, parties, and classes. One point was credited for each activity in which the resident participated once or twice a month; two points were credited for each activity in which the resident participated once a week or more. The individual's participation rate was then divided by the number of activities available in the facility, measured by the POLIF subscale of Social–Recreational Activities. This step controls for differences in the activity program provided by different facilities.

On average, residents participated in about a third of the social–recreational activities organized by their facility. There was considerable variability in participation rate ($SD = 27\%$); the least active residents were involved in no organized activities, the most active participated in all the activities available in their facility. Religious services, bingo, and exercise groups drew many participants, whereas social hours, classes, and discussion groups had somewhat more limited appeal. Some of these facility-organized activities were oriented toward the more impaired residents, who were more likely to participate in them.

On the whole, residents who are more involved in one type of activity are also more likely to be involved in the others. Informal activity in the facility and involvement in community activities were moderately related ($r = .54$), but involvement in these types of informal activities was less strongly related to participation in facility-planned activities (average $r = .30$). In fact, some impaired residents, who were less able to participate in activities in the community, were most involved in organized facility activities.

Next, we examine the personal characteristics and facility features that contribute to involvement in leisure activity. Then we consider whether residents who vary in their functional abilities manifest different patterns of activity involvement in relation to facility characteristics.

DETERMINANTS OF INDIVIDUAL RESIDENTS' ACTIVITY INVOLVEMENT

Personal Characteristics and Activity Involvement

To identify the personal characteristics that are related to activity involvement, we obtained correlations of residents' personal characteristics with their scores on the three activity dimensions after we controlled for the level of care.

Personal characteristics are related to involvement in activities, particularly informal activities. Residents whose functioning is more intact and those with better social resources (higher educational and occupational status and not receiving public assistance) are more likely to engage in informal activities, both within the facility and in the community. Women show a pattern of higher engagement on all three dimensions. Older residents are less likely to go into the community for activities than are younger residents but are more likely to be involved in formal activities. Finally, married residents are more involved than other residents in activities in the community (Table 9.1).

Table 9.1 Partial and multiple correlations from regression analyses predicting individual residents' activity participation from personal and facility factors

	Activity dimensions		
	Activity level (N = 1,416)	Activities in the community (N = 1,404)	Facility-planned activities (N = 729)
LEVEL OF CARE	−.05	−.08**	−.04
PERSONAL FACTORS			
Sex (0 = male, 1 = female)	.17***	.06*	.21***
Age	−.01	−.07**	.13***
Education	.22***	.29***	.00
Occupational level	.11***	.21***	.01
Functional ability	.33***	.32***	−.05
Married	.02	.07**	−.03
Public assistance	−.18***	−.17***	.00
AGGREGATE RESIDENT FACTORS			
Social Resources	.06*	.10***	−.13***
Functional Abilities	.03	.07**	−.15***
Facility score on activity dimension	.13***	.17***	.13***
PERSONAL CONTROL POLICIES			
Provision for Privacy	.08**	.09***	−.04
Policy Choice	−.03	.03	−.18***
Resident Control	.04	.15***	−.09*
Policy Clarity	−.02	.10***	−.10*
SOCIAL CLIMATE			
Cohesion	.16***	.15***	.16***
Conflict	−.18***	−.15***	−.13**
Independence	.15***	.17***	.01
Resident Influence	.13***	.17***	.03
Organization	.19***	.11***	.18***
OVERALL R	*.53****	*.52****	*.46****

Partial correlations control for level of care and personal factors.
*$p < .05$; **$p < .01$; ***$p < .001$.

These findings are generally consistent with earlier research, which shows that health and functional ability are major determinants of activity level, that women are somewhat more involved in activities with family and friends and in a combination of organizations and informal socializing, and that married older people tend to be more active in organizations (Carstensen, 1991; Cutler & Hendricks, 1990; George, 1978; Lawton, 1985a; Russell, 1990). In addition, better-educated older people are more likely to participate in age-related programs and to take an active role in formal organizations (Babchuk et al., 1979; George, 1978); they also spend more time in recreational activity (Lawton et al., 1986–87). Education and occupation may influence activity by developing skills and stimulating interests that carry over into leisure time and by their relationship to economic status and access to resources.

Facility Characteristics and Activity Involvement

As shown in chapter 2, congregate settings vary substantially in their level of resident activity involvement; in chapter 8, we considered some of the facility characteristics that are associated with the overall level of resident activity. Some of these associations may be due to selection processes whereby individuals with a particular activity pattern are more likely to become residents of a setting. Features of the residential setting, such as aggregate resident characteristics, personal control policies, and social climate, also influence the pattern of activity in old age.

We conducted regression analyses similar to those reported in chapter 8 in order to determine whether facility factors are associated with the individual's activity involvement after level of care and relevant personal characteristics are considered. The three activity measures were the dependent variables. We entered level of care and personal variables as the first set of predictors. The different sets of facility characteristics (aggregate resident factors, personal control policies, or social climate indices) were entered as the second set of predictors in the regression equations.

Table 9.1 presents partial and multiple correlations at the stage in the regression analyses when level of care and the personal characteristics have been entered. Thus, the partial correlations for aggregate resident factors, personal control policies, and social climate show their relationship to activity involvement when the individual's personal characteristics are controlled.

Informal Activities

The relationships with facility factors show a similar pattern for both types of informal activities, activities in the facility and community-based activities. Even with level of care and personal characteristics controlled, an individual's informal activity is related to the aggregate characteristics of facility residents (Table 9.1). Where residents as a group have more social resources and are more active, individual residents are more likely to be involved in informal activities within the facility and in the community. In other words, the resident aggregate's impact on individual activity involvement is independent of the individual resident's own background characteristics and functional ability.

Provisions for residents' privacy are also associated with more involvement in informal activities in the facility and in the community. Resident control and policy clarity are associated with more participation in activities in the community. In addition, all the social climate factors are related to residents' involvement in informal activities; specifically, a cohesive, self-directed, and well-organized social climate with low levels of conflict is associated with higher levels of informal activity in the facility and in the community.

These findings are consistent with the results of our regression analyses on the facility level: Independence-oriented facilities with policies emphasizing personal control have a higher aggregate level of resident community involvement (chapter 8; see also Lemke & Moos, 1981). Segal and Aviram (1978) obtained similar findings in residential care facilities serving a psychiatric population; residents in cohesive, independence-oriented, and well-organized programs showed higher levels of activity.

Personal and facility characteristics account for about 27% of the variance in the individual's level of informal activity in the facility and in the community. Shared effects account for about half this variance; personal characteristics account for somewhat more unique variance (8%) than do facility characteristics (5%). The shared variance points to the importance of selection processes whereby individuals with characteristics predisposing them to more activity involvement are likely to reside in settings with features that support such activity. In turn, these settings contribute to the maintenance of high functioning and engagement in activity, both for the individual and for the group as a whole.

Facility-Planned Activities

The individual's use of the organized facility activities is not related to the level of care nor to such individual characteristics as functional abilities and education. However, as noted previously, women residents and older residents are more likely to participate in facility-planned activities. In addition, residents participate more in organized activities in facilities in which the overall level of involvement in such activities is higher and in which aggregate resident functioning and social resources are lower (Table 9.1).

As in our facility-level findings, higher levels of policy choice, resident control, and policy clarity are associated with less participation in facility-planned activities. These findings support the idea that policies emphasizing personal control tend to produce a more individualized pattern of use of the formal activity program.

Three of the social climate factors are related to individuals' participation in facility-planned activities. Residents are more likely to take advantage of these activities when the social climate is cohesive, when residents report little conflict, and when the facility is well organized.

The personal and facility factors considered here account for 21% of the variance in individuals' use of the available facility-planned activities. In contrast to the situation for informal activities, most of the variance (16%) is unique to facility characteristics. The personal factors account for very little unique variance (2%) and share some variance (about 3%)

with facility factors. These results are consistent with the findings at the facility level of analysis (chapter 8); they indicate that facility features can have a significant impact on individual residents' involvement in planned activities.

RESIDENTS' FUNCTIONAL ABILITY AS A MODERATING FACTOR

Functional Ability and Person–Environment Congruence

As we noted in chapter 8, a number of models predict differential sensitivity to environmental conditions based on an individual's functional ability. Adaptation models point to the restricted adaptive capacity of impaired older people and their consequent vulnerability and responsiveness to conditions in their immediate environment. Competence models see functionally able individuals as better able to respond to environmental opportunities and therefore as likely to show greater sensitivity to their environment.

Needs-hierarchy models encompass both of these approaches in that responsiveness to environmental conditions is seen to depend on the particular match between the person and the environment. Features that remove barriers or compensate for the deficits of impaired residents are likely to influence their outcomes more than those of functionally intact residents; environmental features that challenge residents to function autonomously may have more impact on intact than on impaired residents.

A few studies have examined the joint contribution of personal and environmental characteristics to the activity level of older people. For example, Sherman (1974) compared the activity patterns of older people in various types of retirement housing with demographically matched community controls. For the relatively young and healthy residents of a retirement community, the easy availability of activities and recreational facilities made no difference in their activity involvement when compared with their community controls. These individuals appeared to have the necessary economic and functional resources to maintain their activity level whatever the environmental resources. For older residents with significant impairment, however, resources in the immediate environment did seem to contribute to the maintenance of higher activity levels as compared with the matched community controls.

In general, our person–environment matching model assumes that the more competent individuals are likely to be activated by increased opportunities for personal control, whereas personal control policies may be less salient for impaired individuals, who are likely to lack the personal resources to take full advantage of such personal control opportunities.

Thus, policies that promote personal choice and independence (and therefore create a higher level of environmental challenge) should more strongly enhance involvement in informal activities among high-functioning than among low-functioning individuals. In such an environment, the more competent residents are likely to exercise their option to engage in only a few, personally relevant organized activities, so their participation in facility-sponsored activities might be relatively low. Very impaired residents, too, might withdraw somewhat from facility-planned activities in response to the high level of demand for personal initiative, although the environmental challenge might also contribute to the maintenance of their involvement in informal activities.

Supportive interventions, such as activity programming, the presence of other residents who participate in planned activities, and cohesive relationships and good organization may be especially likely to benefit relatively impaired residents. Such conditions create a supportive context, provide important resources, and have relatively little demand quality, so they may have positive effects primarily on impaired residents. Lawton (1989) has predicted, however, that such conditions may provide too much support and may be detrimental to well-functioning residents.

In order to determine whether facility characteristics vary in how they influence activity involvement among residents who differ in functional ability, we carried out additional regression analyses. Based on prior research and theoretical expectations (chapter 8), we focused on whether the influence of aggregate resident factors, personal control policies, and social climate differs depending on the individual's functional ability. In particular, we examined the hypothesis that functionally intact residents benefit more from policies that provide them with personal control and from a cohesive, self-directed social climate than do impaired residents. We were also interested in whether there is any evidence of a detrimental impact on impaired residents when environmental demands are high, as suggested by Lawton's congruence model.

For example, the results in Table 9.1 indicate that residents in settings with an independence-oriented social climate are more likely to be involved in informal activities in the facility. To determine whether this relationship holds more strongly for better-functioning residents, we conducted a separate regression analysis in which we entered individuals' functional ability scores, followed by facility scores on independence, and the interaction term (functional ability $\times$ independence expressed as deviation scores) as predictors of activity involvement. Similar analyses were conducted for the other facility characteristics shown in Table 9.1. The 12 facility features and 3 dependent variables generated 36 regressions; in 23 of these, the interaction term was statistically significant ($p < .05$). In other words, the relationship between activity involvement and the facility characteristic varied with the resident's functional ability.

Functional Ability, Facility Characteristics, and Informal Activities

In Figures 9.2 through 9.5, *high-functioning* refers to residents who are one standard deviation above the mean for this sample on the BIF Functional Ability subscale; these residents can carry out all daily activities without assistance. *Moderate-functioning* refers to residents who are at the mean for this sample; they can carry out most (83%) of their daily activities without assistance. *Low-functioning* residents are those one standard deviation below the mean of the sample; these residents can carry out only about 66% of their daily activities without help.

A general pattern of findings holds for informal activities, both within the facility and in the community. Personal control policies and social climate indices are positively related to informal activity for high-functioning and moderate-functioning residents; for low-functioning residents, these relationships are weaker but remain positive.

Figure 9.2, for example, shows that high-functioning residents are much more likely than low-functioning residents to be involved in informal activities in the facility; this is consistent with the findings shown in Table 9.1. Moreover, involvement in informal facility-based activities generally increases as the facility emphasis on independence increases. In addition, this positive relationship is stronger for intact than for impaired residents. Specifically, high-functioning residents in facilities that are low on independence (one standard deviation below the mean for this sample) partic-

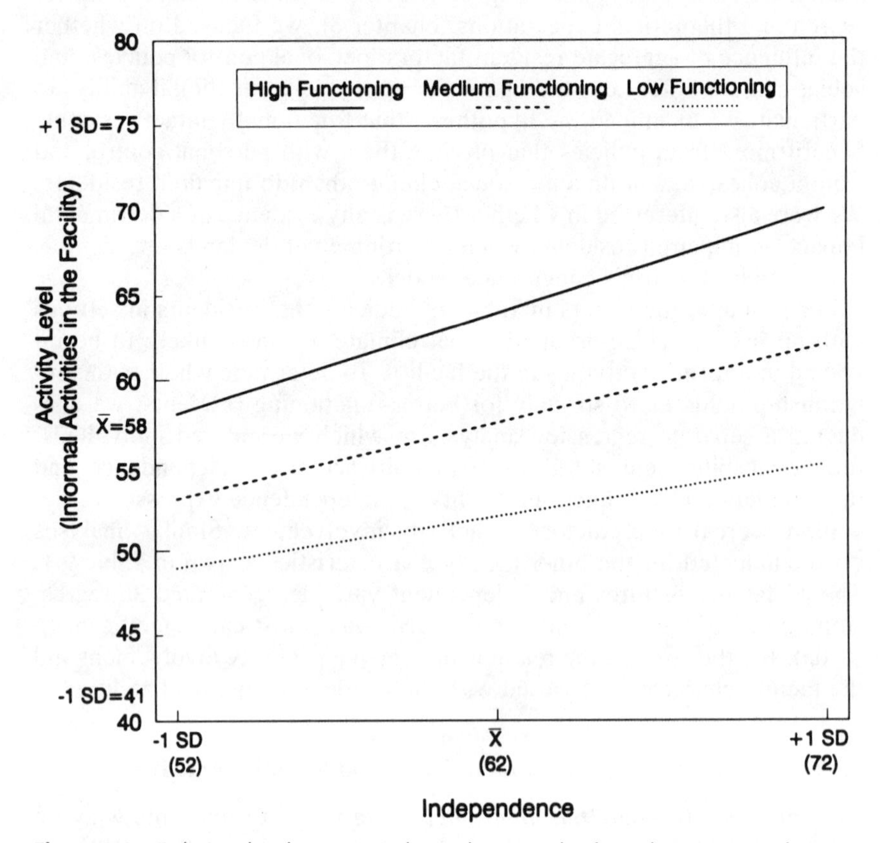

Figure 9.2 Relationship between independence and informal activity involvement for residents of varying levels of functional ability.

ipate in 57% of the informal activities. In facilities high on independence (one standard deviation above the mean), high-functioning residents participate in 71% of these activities. Thus, residents with intact functioning show an increase of about seven percentage points for each one standard deviation increase in independence. As functioning declines, this relationship becomes weaker.

This sample did not include sufficient numbers of very impaired residents to determine whether this relationship reverses to produce a decline in informal facility-based activity for the most-impaired residents. However, our facility-level analyses showed no such decline among very impaired resident groups, that is, among groups of residents who could carry out only about 34% of their daily activities without assistance (chapter 8).

We obtained similar results for the impact of seven other facility characteristics on informal facility-based activities. These characteristics are the three aggregate resident factors (social resources, functional abilities, and activity level), three of four personal control policy measures (privacy, choice, and control), and cohesion. High-functioning residents are more involved in informal activities in facilities whose policies emphasize personal control and whose resident group has high social resources, functional abilities, activity level, and cohesion. For more-impaired residents, involvement in informal facility-based activities shows less variability in relation to these facility factors.

The findings for involvement in activities in the community are comparable: The associations between facility characteristics and community activity involvement are stronger for high-functioning residents. Specifically, such residents tend to participate more in activities in the community when they live in facilities where other residents have more social resources, are more functionally able, and are more involved in community activities; where personal control policies are higher; and where residents report more cohesion, less conflict, and more independence. These facility features appear to be less important in determining the community involvement of impaired residents.

For example, as shown in Figure 9.3, the community involvement of high-functioning residents is strongly related to the facility's provisions for policy choice. This association weakens as the residents' functional ability declines. More specifically, each one standard deviation increase in policy choice is associated with an increase of about seven percentage points in participation in community activities among high-functioning residents and with almost no change among low-functioning residents.

Functional Ability, Facility Characteristics, and Facility-Planned Activities

The pattern of interactions for participation in facility-organized activities differs from the general pattern found for informal activities; participation by impaired residents is more closely related to facility features than is

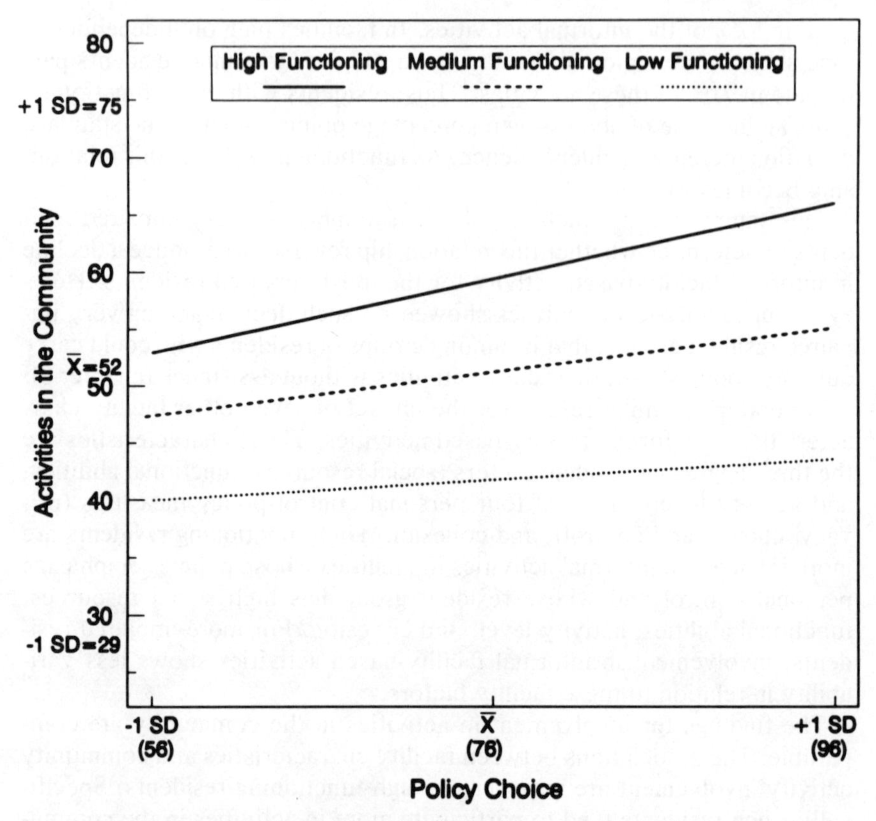

Figure 9.3 Relationship between policy choice and activities in the community for residents of varying levels of functional ability.

participation by high-functioning individuals. We found two variants of this pattern.

Aggregate Participation in Facility-Planned Activities and Social Climate

One variant implies that the more impaired individuals are drawn into facility activities in the presence of environmental resources such as active participation by the resident group as a whole and an emphasis on resident influence and organization. An increase in organization, for example, is related to increased individual participation in facility activities, as reported in Table 9.1. However, as shown in Figure 9.4, this relationship holds strongly for low-functioning residents and weakens as functioning improves. Essentially similar findings hold for resident influence and aggregate participation in organized activities.

Personal Control Policies

A second variant of this pattern involves personal control policies. The analyses reported in Table 9.1 indicate that, on average, individual residents in settings with more policy choice participate in fewer facility-

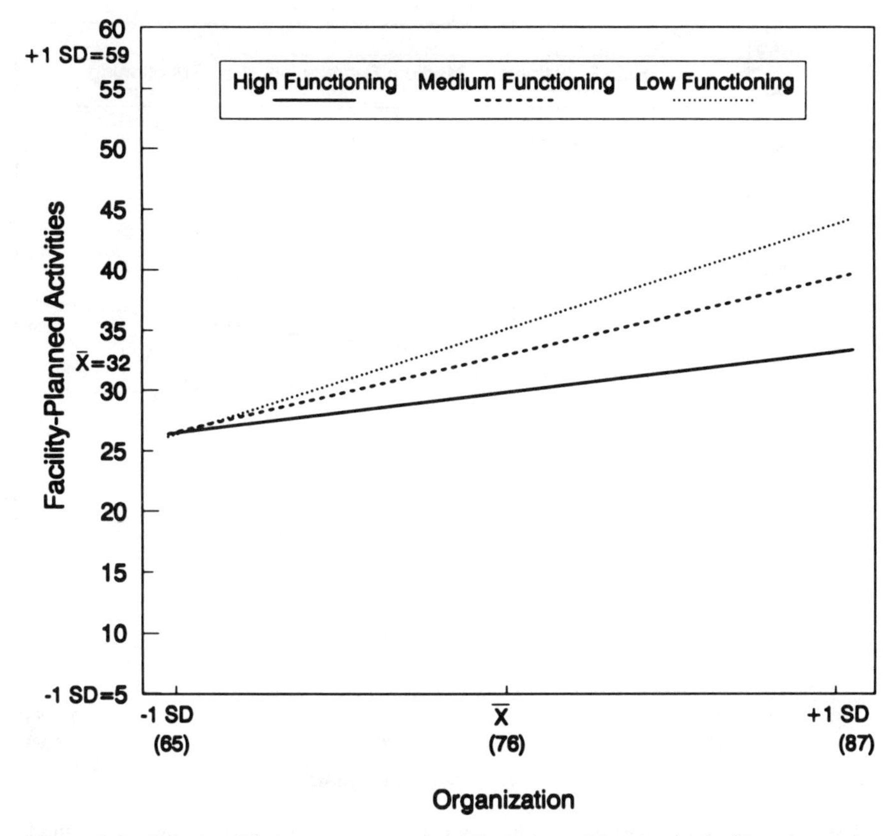

Figure 9.4 Relationship between organization and participation in facility-planned activities for residents of varying levels of functional ability.

planned activities. However, Figure 9.5 shows a more complex relationship. As policy choice increases, high-functioning residents increase their involvement in the facility's planned activities, whereas residents with moderate or low functioning are less likely to participate in such activities. The results are similar for resident control and privacy.

Facilities that are high on personal control policies appear to offer a more diverse program of organized activities, many of which may be geared to the needs of better-functioning residents. Their orientation toward resident choice and control fosters individuality in decisions about activity involvement, even for those residents who might need some staff assistance or encouragement to participate. In addition, as we saw in chapter 8, facilities with a high level of personal control policies tend to have a lower aggregate rate of participation in facility-organized activity. Impaired residents appear to be particularly strongly influenced by this overall level of involvement.

Importantly, however, the impaired residents in these high-choice facilities were at least as involved in informal activities in the facility and in the community as their similarly impaired counterparts in facilities low

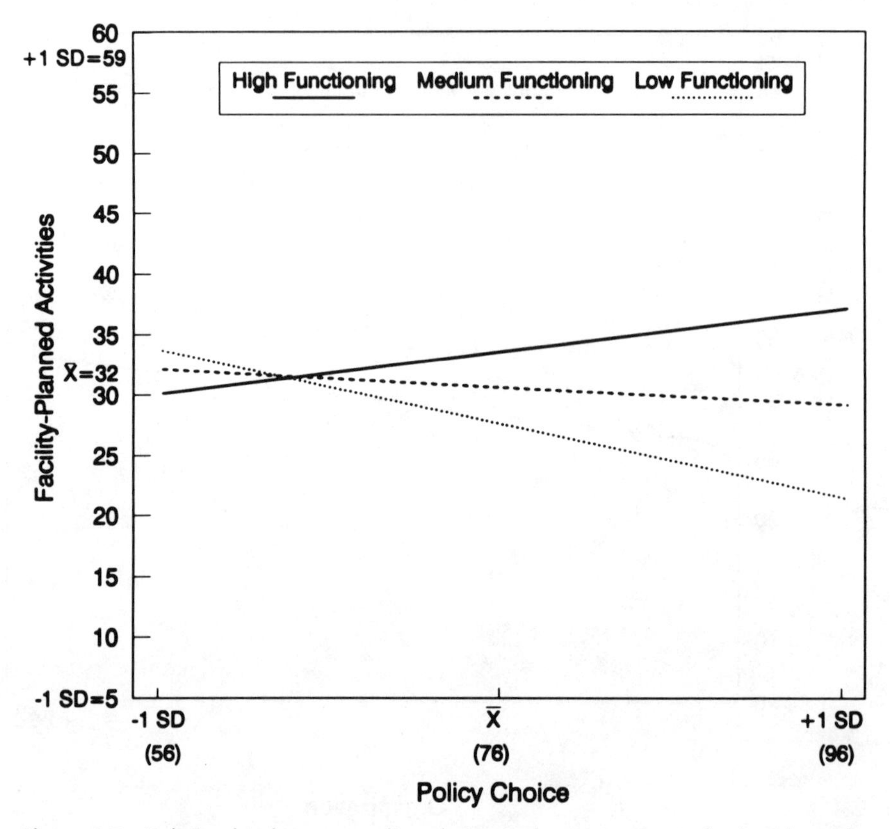

Figure 9.5 Relationship between policy choice and participation in facility-planned activities for residents of varying levels of functional ability.

on policy choice. Thus, in high-choice facilities, the absence of mass-participation activities may lead relatively impaired residents to participate less in the formal activity program but may also help them to maintain their involvement in informal activities and activities in the community.

TOWARD A THEORY OF ACTIVITY INVOLVEMENT

Although researchers sometimes treat all forms of leisure activity as interchangeable, it is useful to distinguish among different kinds of activities. For older people living in a congregate residential setting, it is particularly important to distinguish between informal activities, which can be initiated by the older person, and those that are organized by the facility. These forms of activity have quite different patterns of personal and environmental determinants. In general, involvement in informal activities is related to personal characteristics, to the facility's aggregate resident factors and social climate, and to the mutual selection between individuals and facilities. Involvement in facility-organized activities is related to facility characteristics that draw and push residents into the congregate life of the facility rather than to personal factors or selection processes.

Participation in Informal Activities

Whether it occurs within or outside the facility, informal activity is related to similar sets of personal and facility attributes.

Personal Factors

Consistent with earlier findings (George, 1978; Morgan & Godbey, 1978), older people who are more independent in self-care engage in more independent recreational activity. The same factors that interfere with self-care, such as chronic illness, frailty, mobility impairment, or intellectual confusion, tend to interfere in a similar manner with activity involvement. With increasing age, residents reduce their involvement in community activities; they appear to compensate with increased use of facility-planned activities.

Women tend to engage in more informal leisure activities, as do better-educated older people and those with higher occupational status. In this age cohort, these groups have a well-established pattern of social and recreational activity separate from their work roles. These factors point to the importance of individuals' preexisting, long-standing patterns of activity involvement in determining their independent activity level in congregate settings (Cutler & Hendricks, 1990; Stones & Kozma, 1989).

Facility Factors

Residents are more likely to engage in informal activities if they live in settings where residents as a group come from more privileged backgrounds, have higher functional abilities, and are more active. To some extent, this finding reflects individuals' tendencies to live in facilities in which other residents have a similar social background, but it also appears to reflect the influence of the social milieu on individual functioning. The overall activity level helps define a norm for the setting that in turn appears to influence the individual resident. Observing others in an activity may also stimulate a latent interest.

Policies that support personal control and a social climate that residents see as friendly, high on resident independence and influence, and well-organized also are related to involvement in informal activity. Much of the variance accounted for by these facility factors is shared with personal characteristics. These results point to selection processes whereby older people with characteristics predisposing them to higher activity involvement (intact functioning and more social resources) are more likely to reside in settings with features that support activity. Most of this overlap is probably due to differential selection and retention, but some may occur because settings that promote resident activity contribute to the maintenance of independent functioning. Conversely, a facility with an inactive resident population, more authoritarian policies, poor interpersonal relationships, and disorganization will tend to produce disengagement on the part of individual residents.

The Interplay of Functional Ability and Facility Factors

The individual's functional ability moderates the association between facility characteristics and informal activities. A number of facility features appear to increase the informal activity of functionally intact and moderately impaired residents but to have little or no impact for very impaired residents. Other facility features, such as policy clarity, resident influence, and organization, are related to activity involvement across a range of functional abilities.

Residents who are independent in their daily functioning tend to be more active in informal activities when living in settings where other residents have more social resources, better functional abilities, and greater involvement in informal activities; where policies promote personal control; where relationships are harmonious and resident independence is encouraged. These environmental factors seem to stimulate independent activity for residents who are relatively intact or only moderately impaired. In facilities lacking such resources, these individuals tend to be less active than is expected based on their functional ability alone. However, for very impaired residents, independent activity is generally much lower and less responsive to these environmental factors. Overall, our findings suggest that increasing personal control and self-direction can benefit the informal activity involvement of functionally intact residents and does not seem to be detrimental for the more impaired residents.

Participation in Facility-Planned Activities

In general, participation in facility-planned activities is more closely related to facility characteristics than to personal factors.

Personal Factors

We found that women are more likely to participate in facility activities than men. This is consistent with other findings showing women to be more oriented toward social interaction than are men. Additionally, older residents participate more than do younger ones. Other personal attributes, such as extroversion (Carp, 1985) and mental status (see chapter 10 and Retsinas & Garrity, 1985), also may be related to differential participation.

The absence of demographic characteristics other than gender and age as correlates of organized activity involvement may also reflect the egalitarian nature of facility-planned activity programs. These programs appear to draw residents equally from various social backgrounds and interests. However, the egalitarian nature of the programs may contribute to the impersonal and institutional quality of congregate residential settings by reflecting a failure to accommodate individual differences.

Facility Factors

Residents are more likely to participate in the available facility activities if other residents are involved in these activities and if the group as a whole has fewer social resources and is less functionally independent. Residents also are more likely to be involved in facility-planned activities where settings are more structured and higher on staff control. High participation in such facilities may reflect a general view of residents as relatively passive and subject to staff control.

In addition, the particular mix of activities in a facility may help determine how many residents participate. In some facilities, the formal activities may compensate for the difficulties that the most-impaired residents experience when engaging in informal activities. In other facilities, with minimal staff support and resources, organized activities may attract only the most functionally independent. In yet others, the formal program may include both kinds of activities and have a moderate overall level of participation. In general, these results replicate our earlier finding that facilities with fewer opportunities for personal control tend to promote reliance on facility-planned activity programs and to replace informal, self-initiated activity with facility-organized activities (chapter 8).

The Interplay of Functional Ability and Facility Factors

As with informal activities, residents' functional abilities moderate the association between facility features and participation in planned activities. But, confirming the important difference between informal and facility-organized activities, the pattern of interactions is different. Very impaired residents are more likely to attend the available activities in facilities where the overall participation is greater, where resident influence is higher and the program is seen as well organized, and where the level of personal control is lower. For functionally intact residents, in contrast, their participation in the available activities is higher when the policies support personal choice.

The potential problematic effect of high environmental demands on impaired residents may be diminished by developing a cohesive, well-organized social climate and by maintaining a subgroup of residents who are involved in facility-planned activities. The presence of a stable, well-qualified staff group is also an important environmental resource for more-impaired residents. Accordingly, when staff increase environmental demands by enhancing residents' levels of personal control, they also need to provide a more supportive and structured social climate for very impaired residents. These findings are consistent with the idea that impaired residents need more support in order to function effectively. Cognitively impaired residents, for example, may show less agitation when they are involved in structured activities (Cohen-Mansfield, Marx, & Werner, 1992).

IMPLICATIONS FOR THEORY AND PRACTICE

Our findings support the value of examining personal and facility characteristics in conjunction with one another. Together with the results of prior studies, our findings show that residents' functional ability can have a moderating influence on the association between facility characteristics and resident outcomes. One possible avenue for this moderating influence is individual beliefs about one's ability to control events.

Individuals who believe in their ability to control events are likely to respond more positively to enhanced control. One consequence of reduced functional abilities may be the belief that one can no longer exert such control. For example, Reich and Zautra (1990) found that an intervention to enhance control had a positive effect on mastery and mental health among older adults high on personal control beliefs. Among older adults low in personal control beliefs, however, the control-enhancing intervention had no effect. Instead, an intervention that involved supportive interaction with friendly and empathic individuals had a positive influence on these older adults' mental health. Functionally able residents of group living facilities may respond more positively to personal control policies because they believe in their ability to control everyday events; in contrast, impaired residents may want less control and respond better to support and structure (see also Conway, Vickers, & French, 1992).

Our findings also underline the importance of evaluating facility characteristics in an overall context. Consistent with Lawton's (1989) perspective, policies that enhance personal control may operate more as demands than as resources with respect to impaired residents' participation in facility-planned activities. Together with high expectations for resident functioning and highly active resident group, however, such policies may be associated with more involvement in informal activity among both better-functioning and very impaired residents. Consistent with Sherman's (1974) ideas, moreover, supportive environmental features, such as a well-organized social climate and an experienced staff group, enhance participation in formal activities among impaired residents and have little or no influence on participation among the more competent residents.

When formulating policies and designing interventions, staff can emphasize facility characteristics that are related to informal activity while trying to counteract the potentially detrimental effects of some policies on impaired residents' participation in facility-organized activities. Thus, a cohesive and well-organized social climate that emphasizes resident influence tends to have a positive effect on informal activities, and these social climate factors also increase participation in facility activities among impaired residents.

In general, staff need to help residents maintain a moderate to high level of personal control, which is beneficial for functionally independent residents, and to provide sufficient support and structure to enable impaired residents to achieve their adaptive potential. Intact residents can adapt to an environment with high expectations, but impaired residents need more environmental resources in order to function at their fullest capacity. In the next chapter, we pursue these ideas further by examining how residents of varying mobility and mental status adapted to the process of relocation.

10

Coping with Environmental Change

Studies of relocation of long-term care residents tend to focus on morbidity and mortality rates in response to this stressor. Some research has considered the personal attributes that affect an individual's response to relocation and, to a more limited degree, the environmental changes that can precipitate negative reactions (Baglioni, 1989; Borup, 1983; Coffman, 1981; Grant, Skinkle, & Lipps, 1992; Lieberman & Tobin, 1983). In general terms, relocation provides an opportunity to compare environments and residents' adaptation to them. Along these lines, we used the MEAP to help document the amount and type of change following an intrainstitutional relocation. Here we examine changes in residents' and staff members' use of building areas and patterns of behavior and focus on the association between such changes and differences in the design of the two buildings.

A number of studies have used a longitudinal design to examine changes in resident and staff behavior following relocation. Bourestom and Tars (1974) observed residents' behavior 1 month before and 1 month after either a moderate or radical relocation. Residents who were relocated to an adjacent building where the program and staff remained constant showed little change in activity patterns. Residents who were relocated from a county facility to a large proprietary nursing home in another county, however, exhibited more idle and passive behavior and less involvement in organized activities following the move. The authors viewed these changes as part of a pattern of detachment in response to stress.

In a study of an intrainstitutional relocation, Lawton, Patnaik, and Kleban (1976) examined resident and staff behavior 2 weeks before and 2 weeks after a move. In the new setting, residents restricted their involvement; they spent more time in their bedrooms and less in social spaces, and there was a concomitant increase in passive behavior and a decline in socializing. However, following relocation, residents were more likely to leave their bedroom doors open and showed an increase in self-care behavior, such as organizing their belongings. Staff location patterns paralleled those of residents: Staff members spent more time in resident bedrooms and showed more active behavior after the move. According to the authors, restriction of movement and reduced complexity of behavior are initial adaptive responses to the stress of relocation (Lawton et al., 1976; Patnaik et al., 1974).

Several studies have identified differential effects of relocation on different groups of residents. Focusing on men who were relocated from one building to another in a VA domiciliary, Tesch, Nehrke, and Whitbourne (1989) found declines in residents' morale, number of friends, and attitudes toward their own aging. However, staff ratings of communication showed an improvement, perhaps because residents tried to cope with the stress of relocation by seeking information from and interacting more with the staff. Importantly, sociable residents showed more negative changes after relocation than less-sociable residents did; separation from friends appeared to be more stressful for more-sociable residents.

Nirenberg (1983) assessed the differential effects on high and low mental status residents of a nursing home's relocation from an old, deteriorating facility to a new building. Residents were observed prior to, immediately after, and 3 months after the move. In the period immediately following the move, high mental status residents spent more time in the hallways and less time in their bedrooms; they also showed more active and outgoing behavior. In contrast, low mental status residents exhibited a pattern of withdrawal; they spent more time in their bedrooms and less in the hallways, and they showed more passive behavior and less social interaction. These findings are consistent with the functional competence models of person–environment congruence in that high mental status residents, who have a greater range of adaptive potential, reacted more positively to the challenge of relocation.

Room transfers represent a common type of intrainstitutional relocation; changes in proximity to building areas and in roommates and neighbors cause changes in the physical and social environment. In a study of residents who changed rooms within a long-term care facility, Pruchno and Resch (1988) found higher 1-year mortality among moderately competent residents who moved as compared with similar nonmovers. For high- and low-competence residents, movers and nonmovers showed no such differences in mortality rates. These findings are consistent with the idea that adaptation among moderately competent people is highly dependent on the environment. In contrast, highly competent people have the adaptive capacity to adjust readily to environmental change, and those who are more impaired are less affected by change.

Lyman (1989) focused on the process of staff adaptation when a day-care program for persons with dementia moved from the cramped basement of an old building to a new, specially designed, and more spacious building with safety and surveillance features, nursing and treatment rooms, and additional staff office space. Although these features generally reduced staff work stress, they created some problems. The larger size made it harder for staff to monitor patients and involve them in activities. Moreover, staff support and teamwork declined because the added space and private offices reduced staff members' interaction with one other. Staff coped with these problems by strengthening their communication procedures and altering the structure of the program to maximize staff support and patient self-determination.

Following this line of work, we examine the behavioral patterns of residents and staff in the relocation of a nursing home to a new building, a move that entailed change in some environmental features and stability in others. We consider the short-term impact of the relocation itself and then focus mainly on the long-term impact of the new building on the ecology of the resident population and on particular subgroups. We use a person–environment congruence framework to examine the differential effects of the relocation on individual residents of varied mobility and mental status.

ASSESSING THE IMPACT OF BUILDING DESIGN

Oakview Nursing Home, located in a building not originally designed to accommodate older or disabled residents, was relocated to a modern, architect-designed building. The old building was a two-story, H-shaped structure with a long central corridor connecting two wing corridors (Figure 10.1). Each floor had a nurses' station, some staff offices, and its own dining room and lounge space, all of which were located on the central corridor. Some resident bedrooms were also on the central corridor, and each wing corridor had resident bedrooms and a large lavatory/shower room. An activity room, designated for occupational therapy, was located at the end of one wing corridor on the upper floor. An elevator provided access between the floors. The main entrance to the building was located at the top of a flight of stairs, but a ramp provided wheelchair access to a basement door in back of the building.

The new building is a single-story structure with three distinct 50-bed units. A large patio is at the center of the building; the three units extend outward from three sides of the patio and the communal areas extend from the fourth (Figure 10.2). The main entry, with an automatic door, leads into the communal section, which has staff offices and a conference room, a large dining room, a central lounge, and rooms for occupational and physical therapy. The three units are essentially identical in plan but have different color schemes. Each unit has an open nurses' station, staff

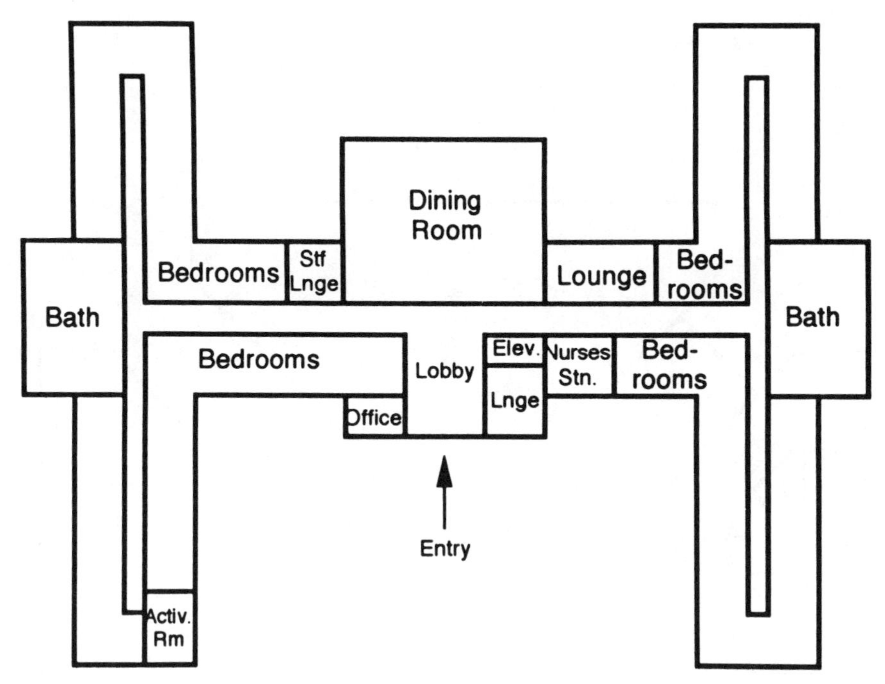

Figure 10.1 Schematic drawing of the old building.

offices, a staff lounge, a small TV lounge, and a congregate bath/shower room. Each resident room also has its own toilet area. Primary colors and graphics are used throughout the new facility, and the main lounge and dining room are carpeted. The large landscaped patio is accessible through an automatic door.

Environmental Assessment

We used the MEAP to help describe Oakview's environmental resources and the changes that accompanied the move. The MEAP was completed prior to the move and 2 months and 10 months after the move.

Because Oakview is a public nursing home, its residents are younger and a higher percentage are men than in the typical nursing home. During the initial observation, the original building, with a capacity of 100 residents, housed 94 men and 4 women. Although the residents' median age was 71 years, their functional abilities and activity level were average when compared with our normative sample. Over 90% of the residents had lived in the facility for more than a year. The staff also was unusually stable; over 70% had worked in the home for more than a year. The staff–patient ratio was 52 full-time patient-care staff members per 100 residents.

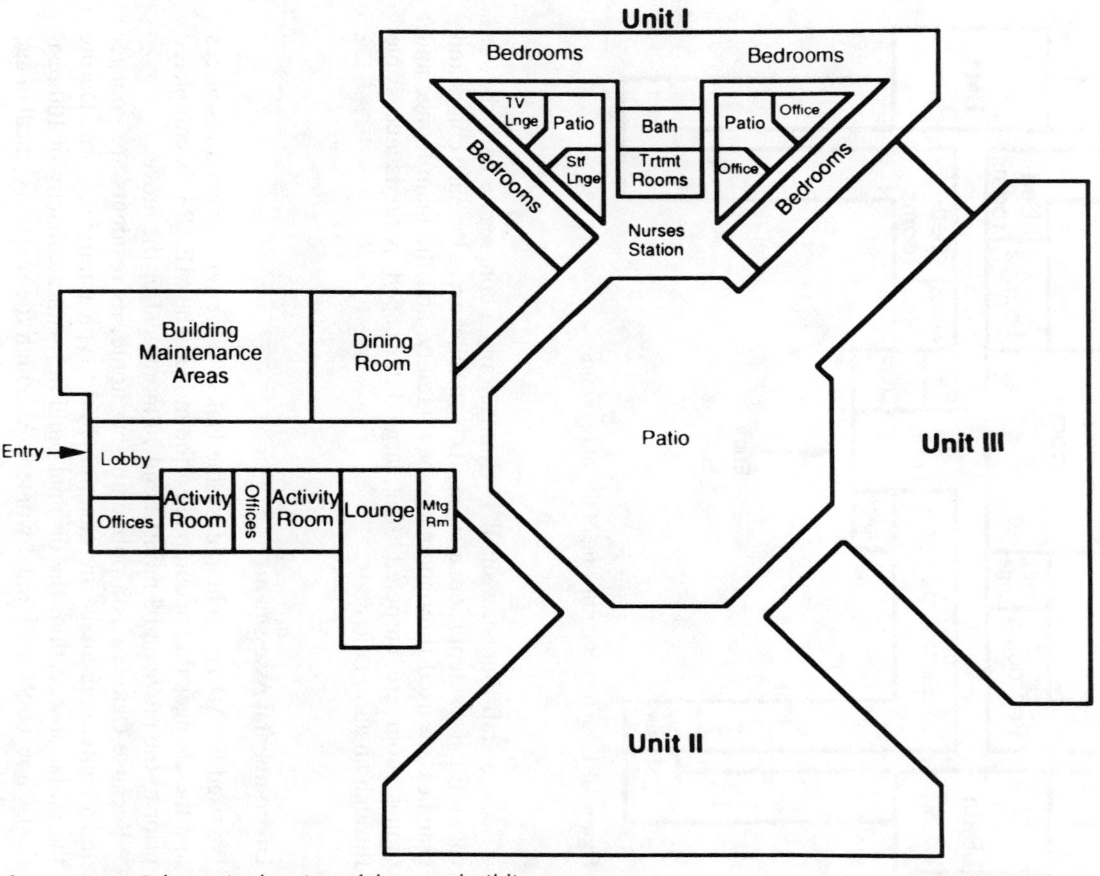

Figure 10.2 Schematic drawing of the new building.

The new building has a capacity of 150 residents. New staff members and residents were added gradually beginning just before the move. Two months after the move, the number of residents rose to 115, and the staffing level increased to 65 patient-care staff per 100 residents. By 10 months after the move, 140 residents were living in the facility, and the staffing ratio dropped to 56 patient-care staff per 100 residents. With the new residents, the proportion of women increased to 7% and the average age of residents dropped to 68 years. Average resident functioning, activity level, and social resources remained essentially unchanged, although new residents were more diverse in social background than the original resident group; the staff also was increasingly heterogeneous.

The relocation produced some important changes in the physical design of the facility (Figure 10.3). The old building offered a variety of social–recreational aids, a relatively spacious physical environment, and easy orientation, but it lacked prosthetic aids and safety features. For example, the hallways were barely wide enough for wheelchairs to pass each other, and bathroom doors opened inward, making access to wheelchairs difficult.

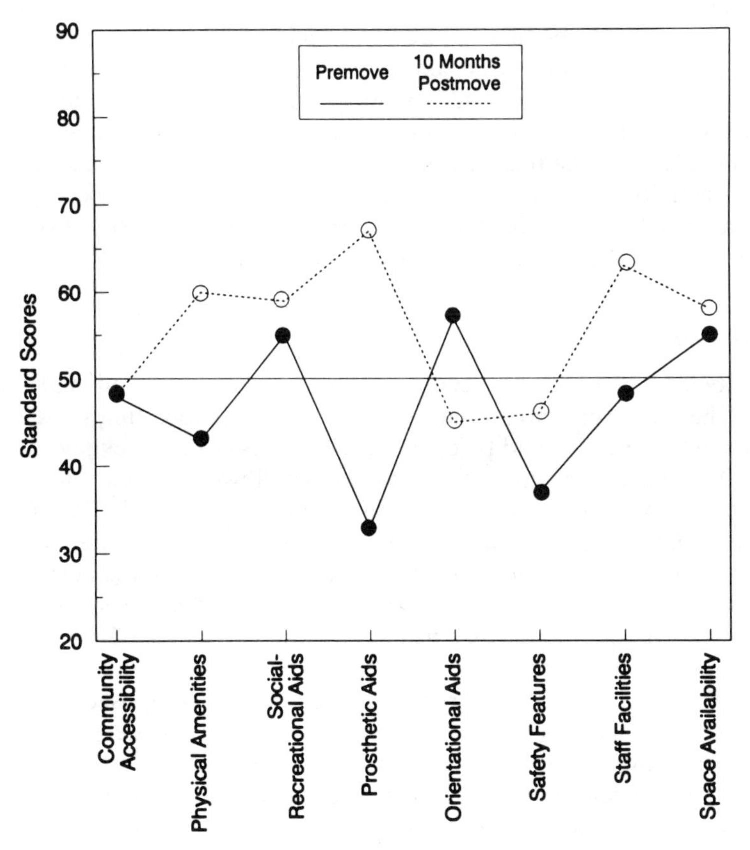

Figure 10.3 Physical features before and 10 months after the move.

As in the old building, the new building provides more space than the average nursing home. One area of marked improvement is the increased provision of prosthetic aids, including wider hallways and automatic doors. The inclusion of call buttons and smoke detectors in residents' rooms constitutes a small but important increase in safety features, and the new building provides more physical amenities. With additional office space, staff facilities increased from about average for a nursing home to above average.

After several months with fewer social–recreational aids because of delays in deliveries, the new setting returned to the previous high level on this dimension. The larger scale and complexity of the new building, however, make it somewhat more difficult for residents to orient themselves. Thus, the new facility provides a richer physical environment in a larger, more complex building. In addition, communal social spaces are a considerable distance from staff caregiving areas.

Prior to the move, Oakview's policies and program were fairly typical of the nursing homes in our normative sample; the exceptions were an extremely low level of policy clarity and, due to its location in a medical center, a very high level of health services. Although the MEAP showed considerable stability over time in facility policies and services, policy choice and policy clarity increased somewhat; in addition, policies shifted toward lower expectations of residents' functional independence but reduced acceptance of disruptive resident behavior. Oakview continued to provide a high level of health services and offered increased assistance in daily living activities and more organized activities. Residents' perceptions of the social climate did not change over the 1-year period even though one-third of the residents were new to the facility.

Behavioral Assessment

We used a behavior-mapping procedure to monitor residents' and staff's behavior. The observers monitored residents and staff four times each day—twice in the morning and twice in the afternoon—each weekday for 2 weeks. At each observation, two observers followed a prearranged route as they walked through the building, and each covered half the building. For each physical location, the observers recorded the persons (coded as *resident, staff,* or *other*) occupying that location and the behavior in which they were engaged. These observation procedures were completed 2 months before the move (Time 1) and 2 months (Time 2) and 10 months (Time 3) after the move (Moos et al., 1984).

We classified residents' behavior into three categories composed of 10 specific types of behaviors. (1) *Passive behavior* consists of sleeping or dozing or being awake but idle. (2) *Active task behavior* is performing self-care activities such as dressing, eating, grooming, or cleaning up; listening to the radio or watching television; engaging in independent activity such as reading, writing, or doing a puzzle; or walking or other forms of locomotion. (3) *Social behavior* includes various forms of interaction, such as conversing, providing or receiving assistance, or engaging in a joint activity. We classified social behavior in terms of the participants as resident–resident interaction, resident–staff interaction, resident–other interaction, and participation in organized facility activities.

Staff behavior was classified into four categories. (1) *Active task behavior* covers activity such as writing reports, cleaning, or preparing medications. (2) *Resident-oriented behavior* consists of conversing with residents, providing assistance, or engaging in a joint activity. (3) *Staff-oriented behavior* consists of conversing, assistance, meetings, or joint activity between staff members. (4) *Nonwork behavior* consists of behavior such as sitting idle, watching television, or eating.

Adaptations in Resident and Staff Behavior

When a residential facility is relocated, residents and staff members try to cope with the new setting and to adapt to the challenges it presents. Salient environmental factors include the physical design and scale of a building and the layout of public and private space. The residents' needs for privacy, social contact, and stimulation are important, as is their functional impairment. Policy and program changes must also be considered. Finally, the way in which staff manage space is a function of their responsibility for resident care and their own needs for order and privacy.

Given these considerations, we thought that the larger size and more complex layout of the new building would produce some disorientation among residents and consequently would cause them to stay closer to their own rooms. In addition, we expected residents to spend time in areas of the building that would give them contact with staff and a view of ongoing activities, such as in the hallways around the nurses' stations. Because the large communal social spaces in the new building are distant from the bedrooms and isolated from unit activity, we expected residents to use these communal spaces less than they did in the old building. Furthermore, it seemed likely that the location, small size, and limited accessibility of the unit lounges would result in low use of these social spaces as well. Finally, we expected that staff–resident interaction might increase somewhat as a result of the higher staffing level and residents' attempts to cope with change.

To examine these ideas, we summed the daily results, each including four observational sweeps, and kept separate each of the 2 weeks of each observation period. We converted the daily totals to percentages based on the number of residents or staff available for observation, and conducted analyses of variance separately for residents and staff. We tested the significance of the differences among the three observation periods (the main effect) against the week-to-week variations in behavior pooled over the three periods (residual variance). Significant results indicate that the variation among the three observation periods is greater than naturally occurring week-to-week fluctuations in behavior.

Residents' and Staff's Spatial Location

Both before and after the move, the observers most frequently found residents in the bedrooms (Figure 10.4). According to policy, staff got residents dressed and out of their beds each day, so these findings represent not only the most physically handicapped residents but also the more competent residents who used their rooms for active task behavior. Bedrooms were the locale for the majority of self-care and independent activity as well as for daytime sleeping. Ten months after the move, bedroom use was significantly higher than during the two earlier observation periods.

The rest of the spatial locations shown in Figure 10.4 are ordered roughly in terms of their proximity to residents' rooms. One striking con-

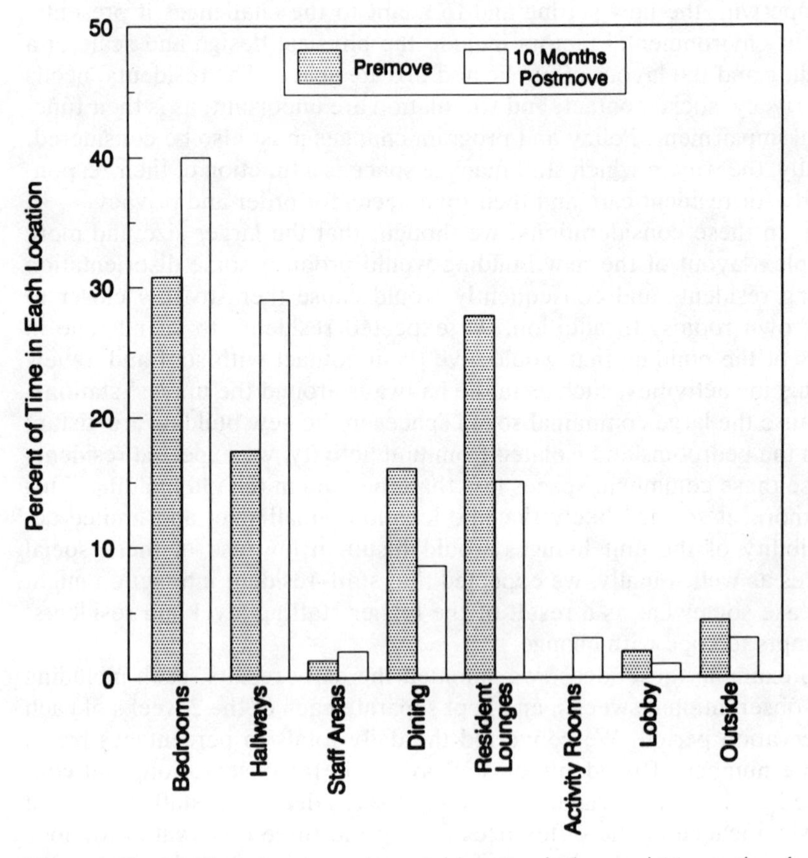

Figure 10.4 Resident time in each spatial location before and 10 months after the move.

trast is the greater use of the hallways in the new building. Although the amount of time residents spent at the nurses' stations and other staff areas remained relatively constant, the dining room and lounge areas of the new building were used much less compared with the old building. The exception to this pattern of reduced use of social spaces occurred in the activity rooms designated for occupational or physical therapy; in these locations, observers noted increases from only trace use initially to just over 3% 10 months after the move.

Changes in staff's spatial location patterns generally mirrored those shown by residents. In the old building, many bedrooms were distant from the nurses' stations, and staff moved the less-able residents to the social areas and nurses' stations for ease of access. Distances between the nurses' station and bedrooms are much less in the new building, and staff can monitor residents and provide care to them in their rooms. Accordingly, staff spent more time in the bedrooms in the new building (an increase from 8% to 13%). Staff used the dining room, lounges, and lobby less in the new building (an overall decrease from 25% to 11%) and used the activity rooms more (up from trace amounts to 3%).

Residents' and Staff's Behavior Patterns

As shown in Table 10.1, residents spent a little over half their time sleeping or idle (awake but disengaged) at all three observation periods and about 45% of their time either in active task behavior or in social behav-

Table 10.1 Mean percent of resident behavior in each behavior category

Type of behavior	2 months premove	2 months postmove	10 months postmove
PASSIVE	*59.4*	*55.5*	*54.1*
Sleep	21.5	17.4	21.8
Idle	37.9	38.1	32.3
ACTIVE TASK	*28.9*	*27.6*	*30.4*
Self-care	4.0	4.1	5.8
Radio/TV	7.7	8.5	9.0
Independent activity	7.5	5.4	6.9
Locomotion	9.7	9.6	8.7
SOCIAL	*13.0*	*18.6*	*19.0*
Interaction with resident	1.3	0.8	2.4
Interaction with staff	5.2	6.8	7.5
Interaction with others	0.6	0.4	1.2
Organized activity	5.9	10.6	7.9

Differences between the three time periods are significant ($p < .05$) for all indices except radio/TV and locomotion. Columns sum to more than 100% because some residents were simultaneously engaged in more than one type of behavior.

ior. Between the initial and postmove observations, the frequency of social behavior increased somewhat, and the amount of passive behavior decreased. Two months after the move, independent activity was significantly though temporarily lower, as was the amount of time spent in napping. In their place, residents were more involved in interaction with staff and in organized activity. Residents may have responded to the uncertainty connected with the relocation by seeking more contact with staff, a pattern observed by Tesch and her colleagues (1989).

Focusing on their longer-term adjustment to the new building, we found that residents showed a higher level of self-care activity 10 months after the move. The level of social behavior also consistently exceeded that found in the old building prior to the move. Specifically, behaviors that involve direct interaction among residents increased somewhat, but they accounted for only about 1% to 3% of the residents' time. Interactions between residents and staff were somewhat more frequent in the new building, but they still comprised only between 6% and 8% of the residents' time. Interaction with people from outside the facility was also higher 10 months after the move than it was before. Finally, resident participation in organized activities almost doubled from before to 2 months after the move but then decreased somewhat in the subsequent 8 months.

Staff behavior broke down as follows: Between 35% and 40% was active task behavior, 20% to 25% was resident-oriented behavior, and 30% to 35% was staff-oriented behavior (Moos et al., 1984). Less than 10% was classified as nonwork behavior. Overall, these rates are generally consistent with other observational studies of staff behavior (e.g., Burgio et al., 1990). Consistent with the increase in policy clarity, staff spent more time in meetings after the move, in part due to the need to coordinate new programs. In addition, staff temporarily increased their participation with residents in organized facility activities 2 months after the move. The staff involvement in organized activities likely accounts for the temporary increase in resident involvement. Otherwise, staff behavior patterns remained highly stable over time.

Spatial and Behavioral Accommodations as Coping Responses

Overall, residents spent more time in the main halls near the nurses' stations in the new building and less time in the dining room, lounge, and lobby areas. Ten months after the move, residents were also more likely than prior to the move to spend time in their bedrooms. We have noted some possible ways in which residents' and staff's needs and coping patterns may help to explain these accommodations to the physical design and larger scale of the new facility.

In contrast with the changes in residents' spatial location patterns, their behavior patterns were relatively stable. The stability in behavior patterns, as well as in resident and staff perceptions of the social environment, point to the inertia of large social systems and to the active efforts of individuals within these systems to maintain stability in the face of change.

There were modest increases in the extent to which residents interacted with each other, with visitors and volunteers, and with staff members as well as in the amount of time they spent in organized facility activities. These changes correspond with increases in program activities and in the attractiveness and openness of the facility to outsiders. As a consequence, residents were less likely to be idle in the new building. Some of these changes were likely due to the behavior of new residents as they were drawn into the social life of the facility.

Transforming Space to Help Monitor Residents

Lawton (1981), who emphasized the importance of providing readily accessible social spaces, pointed out that residents are strongly attracted to areas of high activity. In the present example, the layout of the old building facilitated contact with staff outside of personal spaces by locating the communal social areas near the nurses' stations and at the heart of the building near the entrance and elevator. In the new building, unit lounges are on the back corridor and the main lounge is in a separate wing, removed from proximity to staff work areas.

Staff and residents responded to the building layout by transforming into lounge areas the wide corridors that connect the units. Staff placed chairs and ashtrays along one side of the corridor and decorated the walls. Because of space constraints, these chairs were arranged in a line or row rather than in conversational groupings. Although this arrangement did not encourage sustained social interaction, it did let staff monitor residents and gave residents a view of corridor traffic, the activity at the nurses' stations, and the central patio.

Transforming Space to Provide Staff Privacy

Caregiving requires energy and motivation from staff, who need some time away from residents to cope with the stressors of the job. The new building accommodates this need by providing more adequate staff areas. Staff chose the lounge nearest the nurses' station as the staff lounge, suggesting that they, too, need an area on the edge of activities. This location let staff monitor activities without participating actively. As a consequence, staff shifted in the spatial location of their charting activities and informal interaction between staff members. Ten months after the move, staff spent less time at the nurses' stations (a decrease from 22% to 16%) and more in other staff areas such as the staff lounge (an increase from 24% to 36%).

The Interplay Between Physical Design and Programming

We have focused on changes in the physical environment, but changes in policy and program factors are also important. For example, in the old building, the dining rooms also served as lounges and were equipped with televisions, but this was not the case in the new building, whose dining room was kept locked except during mealtimes and the afternoon social hour. Although the new dining room was more attractive, residents used it less.

The new activity rooms for occupational and physical therapy offer an example of how program changes can compensate for design limitations. The activity room in the old building was used infrequently, in part because of its location at the end of a wing corridor. In contrast to the reduced use of the social areas in which no specific programming occurred, the use of the activity rooms showed a small but significant rise. The occupational and physical therapists who staffed these rooms tried to draw residents into activities outside their units and were having an impact 10 months after the move.

Because any relocation involves simultaneous change in many environmental factors, it is difficult to specify the factors that are necessary or sufficient causes of behavioral adaptations. We have focused on shifts in physical design but have also emphasized the need to consider changes in facility policies, services, and social climate. The relocation studied here entailed the move of an entire group of residents and staff into a new building with a pleasant and spacious physical design, where residents generally chose their room assignments and were placed with familiar peers and staff. These factors likely contributed to the absence of detrimental impacts of the move. Similarly, Mirotznik and Ruskin (1985) found better postrelocation morale among nursing home residents who were moved to a new facility designed specifically for their needs.

A PERSON–ENVIRONMENT CONGRUENCE MODEL OF BEHAVIOR CHANGE

From an interactionist perspective, such as the model of person–environment congruence described in chapter 1 and elaborated in chapter 9, the impact of the residential environment may vary from one individual to another (Nirenberg, 1983; Tesch et al., 1989). Our model focuses on the relationship between personal characteristics, environmental characteristics, and adaptation. Residential relocation is an instance of environmental intervention and offers an opportunity to examine these relationships. We use the congruence framework here to consider the differential impact of the relocation on the behavior of residents who vary in mobility and mental status.

Predicted Interaction of Aids to Mobility and Residents' Mobility Status

Figure 10.5 gives an example of predictions that can be derived from a person–environment congruence model for the connection between an individual's mobility status and the environmental factors that act as a re-

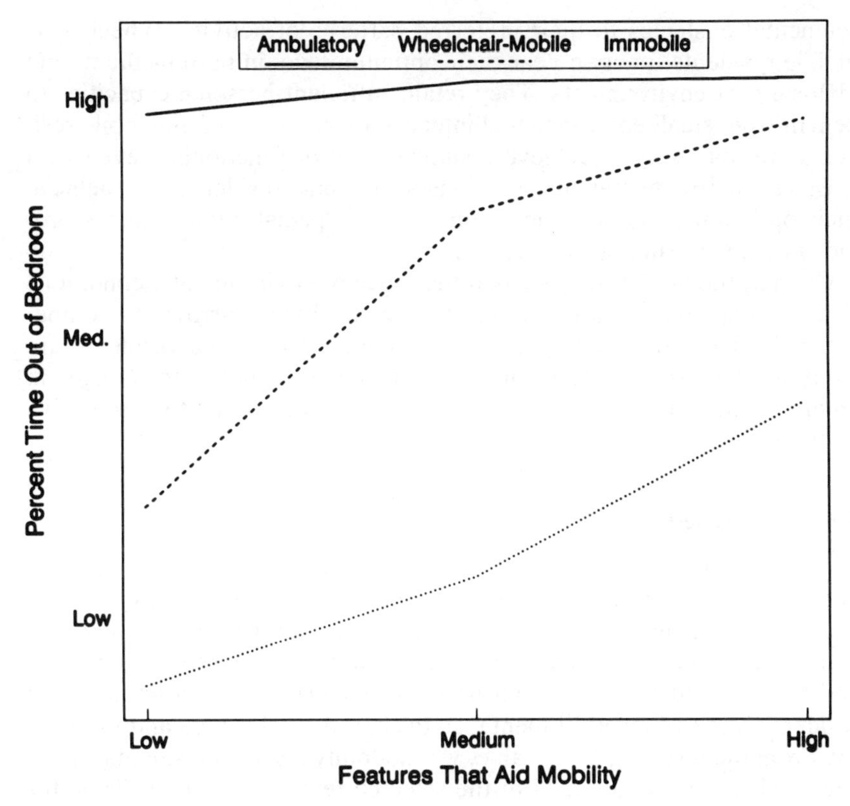

Figure 10.5 The predicted influence of support for access and mobility on residents who vary in mobility status.

source or demand. Here, on the horizontal axis, we plot features that aid residents' mobility. On the vertical axis, we plot an index of activity level, space utilization, which is measured by the percent of time residents spend out of their bedrooms. The graph shows separate predictions for ambulatory, wheelchair-mobile, and immobile residents.

Three hypothetical levels of support are shown on the horizontal axis. A facility with low support would have various barriers to wheelchairs, such as stairs and narrow entries, and would provide few prosthetic aids for residents with impaired mobility. Facilities with medium support would be free of physical barriers to wheelchairs and provide some prosthetic aids, such as handrails, to help frail ambulatory residents but would offer only limited physical and staff resources to help immobile residents. High-support facilities would be barrier-free, provide assistive features, and would provide sufficient staff resources so all residents could leave their rooms and spend time in the social areas.

We expect ambulatory residents, compared with wheelchair-mobile and immobile residents, to function near their adaptive level in all three settings because they have the personal resources to overcome environmental obstacles to their preferred patterns of activity. Wheelchair-mobile residents function near their optimum level in both medium- and high-support environments. They retain sufficient personal capacities to benefit from small environmental interventions. However, immobile residents are not likely to achieve a similar level of functioning, even with massive environmental interventions. Immobile residents can achieve their optimum functioning level only in the special circumstances provided by a high-support facility.

Overall, the model suggests that the range of environmental conditions that is compatible with optimum adaptive behavior narrows as competence declines. It also implies that moderately impaired residents are likely to show the greatest range of behavior in response to changes in environmental conditions and are particularly influenced by supportive features.

Behavioral Assessment

To test our hypothesis, we studied subsamples of residents. The head nurses classified residents into three mobility groups: ambulatory, wheelchair-mobile (able to transfer to a wheelchair and get around independently), and immobile (those requiring assistance in order to transfer from bed to chair and to move about in a wheelchair). Nurses also classified residents into two levels of mental status based on alertness and rationality. From the resulting 2×3 table, we randomly selected residents within each cell for observation; 17 of the selected residents were available for follow-up. Of these residents, 7 were ambulatory, 6 were wheelchair-mobile, and 4 were immobile. The head nurses had identified 7 of these residents as having intact mental functioning and 10 as intellectually impaired. Mobility and mental status were not related for this sample.

Observers monitored these residents for 5-minute segments during four time periods: morning, early afternoon, late afternoon, and evening. Residents were observed for 10 days for each of the daytime periods and on 8 evenings, so the total observation time was 190 minutes per resident. During the 5-minute segments, the observer made a continuous record of the individual's location and behavior.

The categories of behaviors were the same as those used in the observation of the resident group as a whole: passive behavior (sleeping or idle), active task behavior (self-care, radio or television use, independent activity, or locomotion), and social behavior (resident–resident, resident–staff, resident–other, or organized activity). We summarized observations by tabulating the number of minutes during which the individual engaged in each behavior; then we computed percentages by dividing the number of minutes in each behavior by the total number of behaviors observed. We followed a similar procedure for spatial locations.

We conducted analyses of variance to compare resident subgroups. As before, we tested the significance of the differences among the three observation periods against the week-to-week variations in behavior pooled over the three periods. Because the results from the assessments 2 months and 10 months after the move were similar, we present only the long-term changes (Lemke & Moos, 1984).

Residents' Mobility Status and Physical Accessibility

The person–environment congruence model predicts that some of the changes resulting from the move should differentially affect residents in the three mobility groups. For example, the increase in prosthetic aids in the new building (a shift from low support to medium support in Figure 10.5) should have the most impact on residents with impaired mobility. With physical barriers to access removed from the new building, factors other than mobility status assume importance in determining how an individual uses the building. Enhanced by staff assistance in transporting immobile residents, the improved accessibility of the new building could have marked impact on the locations used by immobile residents. However, without such assistance, only the wheelchair-mobile residents should be affected by this particular change.

The floor plans of the two buildings (Figures 10.1 and 10.2) highlight another significant change. In both buildings, most of the communal social spaces are near the entrance and adjacent to the main corridors. However, whereas the old building placed staff offices and the nurses' stations near these social areas, the new building separates caregiving and planned social areas. As noted earlier, residents as a group spent more time in the hallways and bedrooms and less time in the dining room and lounges following the move (Figure 10.4). The immobile residents, who are most in need of staff contact, should be affected most by this aspect of the new design. In contrast, the more mobile residents might be drawn to the social areas, which are somewhat more attractive in the new building, despite the social areas' increased distance from staff.

Our findings showed that mobility status did affect where and how individuals spent their time in the old building and was related to changes following the move. The old building, with its physical barriers to mobility, tended to segregate wheelchair-mobile and immobile residents from ambulatory residents (as in low-support facilities in Figure 10.5). In the new building, ambulatory and wheelchair-mobile residents overlapped more in their space utilization (as in medium-support environments in Figure 10.5). Both before and after the move, ambulatory and wheelchair-mobile residents were similar in their activities and more active than the immobile residents.

Ambulatory Residents

Prior to the move, ambulatory residents used all areas of the building except the activity room; they spent about a third of their time in their bedrooms; a third of their time in the lounges; and the remainder in the

Table 10.2 Mean percent of resident behavior in each behavior category (by mobility status)

Type of behavior	Ambulatory ($N = 7$)		Wheelchair-mobile ($N = 6$)		Immobile ($N = 4$)	
	Premove	10 months postmove	Premove	10 months postmove	Premove	10 months postmove
PASSIVE	*44.7*	*46.8*	*27.8*	*32.0*	*60.4*	*66.3*
Sleep	10.3	12.5	7.0	9.0	15.3	25.9*
Idle	34.4	34.3	20.8	23.0	45.1	40.4
ACTIVE TASK	*39.5*	*35.1*	*53.4*	*50.0*	*16.3*	*10.3*
Self-care	9.6	8.8	13.6	13.8	0.7	0.5
Radio/TV	17.2	10.8*	18.8	12.6*	13.3	7.7*
Independent activity	3.6	1.6*	10.3	8.8	0.8	0.7
Locomotion	9.1	13.9*	10.7	14.8*	1.5	1.4
SOCIAL	*15.7*	*18.1*	*18.8*	*18.1*	*23.4*	*23.5*
Interaction with resident	3.9	5.0	3.4	3.0	0.2	0.6
Interaction with staff	5.4	7.0	7.7	8.3	19.5	20.2
Interaction with others	0.3	0.2	0.7	0.1	0.0	0.2
Organized activity	6.1	6.0	7.0	6.7	3.7	2.5

Percentages are based on total number of behaviors observed.

*Differences between premove and postmove levels are significant, $p < .05$.

dining room, hallways, or outside. Compared with the other groups, they were intermediate in terms of both their passive and active task behavior; they were also lowest in social behavior, due mainly to their low level of interaction with staff (Table 10.2).

Compared with wheelchair-mobile residents, ambulatory residents had particularly low levels of self-care and independent activity. Several factors may account for their initially low activity level. Many of the ambulatory residents entered residential care because of psychiatric diagnoses and had long histories of living in an institutional setting. Although they had the physical capacity for most activities, even some of those who were alert and well oriented had lost or never developed a pattern of active use of leisure time. In addition, some ambulatory residents were involved in activities outside the nursing home and used it as a place to rest between these forays. Because the residents were observed only in the nursing home and its grounds, the observers did not record such activity.

After the move, ambulatory residents made some changes in space utilization; most striking, they spent more time out of their bedrooms (Figure 10.6). The amount of time they spent in the hallways, which were used

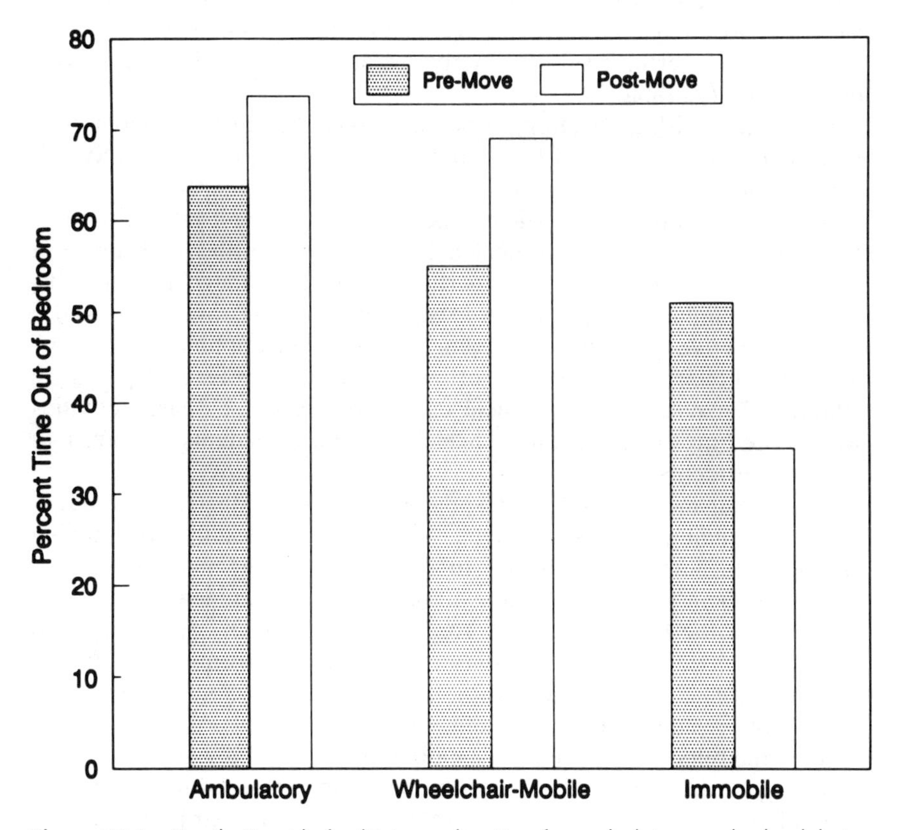

Figure 10.6 Prosthetic aids facilitate exploration for ambulatory and wheelchair-mobile but not for immobile residents.

as lounge areas in the new building, increased markedly (from 8% to 35%). To compensate, the ambulatory residents reduced their use of the more distant lounge and dining room. In response to the dispersed layout of the new building, they also spent more time in transit (Table 10.2). Because these residents spent less time in areas with a television set, they watched television less frequently, which suggests that their viewing was relatively passive.

Ten months after the move, the ambulatory residents' already low level of independent activity fell below 2%, which may have been an early indicator of a decline in health status that had affected their cognitive functioning and resulting activity level but had not yet affected their ability to walk. A number of researchers have found that such cognitive declines can presage illness or death (Kastenbaum, 1985).

Wheelchair-Mobile Residents

Wheelchair-mobile residents were more restricted in their use of the old building than were ambulatory residents. For example, they spent almost as much time in their bedrooms as did the immobile residents (Figure 10.6). In spite of their spatial restriction, wheelchair-mobile residents were the most active group. They were least likely to sleep or be idle and most likely to perform self-care activities and to engage in independent leisure activity (Table 10.2).

In contrast with ambulatory residents, nonambulatory residents were more likely to be admitted to the nursing home with only a medical diagnosis and to have an established pattern of leisure activities. Unlike immobile residents, wheelchair-mobile residents retained sufficient physical capacity to transfer to a wheelchair, propel themselves, and carry out other activities independently. In particular, those with high mental status strove to maintain independence in self-care activities and continued a pattern of active involvement (reading, writing, doing hobbies); these residents used their bedrooms as quiet, private space for these purposes. In addition, because of lost physical capacities, many self-care activities were more time-consuming for wheelchair-mobile residents than for ambulatory residents.

Wheelchair-mobile residents responded to the more dispersed layout of the new building and its improved wheelchair accessibility by moving around even more (up from 11% to 15%). They spent much more time outside of their bedrooms (Figure 10.6). The most prominent change was their use of the hallways, which rose from 18% to 33%. But they also used the outside and special activity rooms more (up from trace amounts to 8%). These changes increased the similarity of wheelchair-mobile and ambulatory residents' use of building areas.

After the relocation, wheelchair-mobile residents, like ambulatory residents, spent less time in rooms with televisions and less time using this form of entertainment (Table 10.2). In other respects, their pattern of behavior was stable. They continued to spend a little less than a third of their time inactive and to be the most likely of any residents to engage in self-care and independent activity. Although wheelchair-mobile residents were more likely than were ambulatory residents to be observed in independent activity, they were about equally likely to be involved in organized facility activity, and these levels were maintained in the new building. Their interactions with staff and other residents did not change in character or frequency.

Immobile Residents

Many of the differences in activity that distinguish the immobile residents from the others can be traced to their physical status. Initially, the immobile residents' behavior was limited in range and consisted mainly of passive behavior, watching television or listening to the radio, and interactions with staff (Table 10.2). Their spatial range also was restricted.

Although the new building had improved wheelchair accessibility, and immobile residents were observed at least once in every area of the new building, they were not more likely to be moved around than they had been in the old building (locomotion was stable at about 1.5%). In addition, in the new building, bedrooms are in closer proximity to staff areas, whereas communal areas are more distant; thus, staff were able to monitor residents in their bedrooms more easily than in the central social areas. Accordingly, the time immobile residents spent out of their bedrooms declined from about 51% to 34% (Figure 10.6). They spent only 8% of their time outside the bedrooms or unit hallways and were much less frequently located in a room with a television and consequently watched less television (Table 10.2). Isolated in their rooms and with less access to television, these residents had reduced stimulation and slept more; sleeping increased from 15% to 26%, and total passive behavior constituted 66% of their time.

Physical and Social Barriers to Mobility

As expected, the improvement in prosthetic aids in the new building affected wheelchair-mobile residents most. Wheelchair-mobile residents were not dependent on others for mobility, and physical barriers to wheelchair access were minimal in the new building. Thus, these residents were able to respond to the more dispersed layout of the new building by spending more time moving around, as did the ambulatory residents. Their pattern of space utilization shifted so as to be nearly indistinguishable from that of ambulatory residents. The new building was thus more egalitarian with respect to mobility status than the old building had been.

Immobile residents were affected in somewhat contradictory ways by the move. Although the reduction in physical barriers increased spatial range for residents who used wheelchairs, immobile residents depend on others for mobility. Thus, their use of space was affected not only by physical accessibility but also by social factors, such as the availability of assistance and their need for staff monitoring.

These social barriers are evident in the differences between wheelchair-mobile and immobile residents. In the old building, for example, immobile residents spent even less time in the lounges than did the wheelchair-mobile residents. Similarly, the lobby and staff areas were never used by immobile residents, although they were used occasionally by others. (See Stephens, Kinney, & McNeer, 1986, for a discussion of other barriers to spatial integration of different ability groups.)

Residents' Mental Status and Ease of Orientation

According to the congruence model, differences in residents' mental status should be associated with differences in their use of building areas and in their behavior patterns. For example, the larger scale and complexity of the new building, including the many 45-degree angles and triangular arrangement of the units, might stimulate increased exploration for high mental status residents but inhibit movement outside a restricted territory for low mental status residents. The inhibitory effect of increased complexity should be especially strong when there is no accompanying increase in orientational aids, as was the case in the new building.

The social spaces in the new building are less accessible both because of the larger scale and complexity of the design and because of their relatively distant location. These factors should have little or no effect on high mental status residents but should result in less use of these spaces by mentally impaired residents, especially when there is no additional staff support to escort residents to activities or social areas. Intellectually impaired residents are likely to function near their optimal level only in a small building with a simple design and in which communal areas are close to caregiving areas and staff are easily accessible. As in the model predicting outcomes for residents with different levels of mobility (Figure 10.5), this application of the model predicts that the more intact residents can function near their adaptive level in a variety of settings, whereas impaired residents are able to achieve their best functioning only in a restricted range of environments.

High Mental Status Residents

In the old building, the high mental status residents spent about half their time in their bedrooms, which served as the locale for most self-care and independent activity. These residents were able to generate stimulation for themselves and did not require much staff monitoring; these factors help account for their staying in their rooms more than did the low mental status residents (52% vs. 35%). Similarly, residents who were more alert ventured outside the building more often than did other residents (7% vs. less than 1%). Like bedrooms, these outside areas could not be easily monitored by staff.

In the new building, high mental status residents spent much less time in their bedrooms, decreasing to a level similar to low mental status residents' (about 36%). Instead, they spent more time in the hallways (up from 15% to 24%), which were being used as social spaces. Unlike the low mental status residents, high mental status residents used the new lounges more often than they did the old (up from 10% to 14%); again, this difference suggests that the need for proximity to staff is not as important in determining spatial location for this group as for the residents who are less alert. After the move, these areas may have attracted more high mental status residents precisely because they were more likely to encounter other intellectually intact residents there. Moreover, the increased attractiveness of the lounges and greater complexity of the new building design may have stimulated the high mental status residents to explore more.

Following the move, high mental status residents showed increased locomotion, decreased television viewing, and increased interaction with staff. Their passive behavior also increased from 20% to 25%, which may reflect the downward course of chronic diseases among these initially more intellectually intact residents. However, this change may also have been a reaction to an additional year of living in an institution or to some features of the new building. For example, by leaving their bedrooms to spend more time in the lounge and hallways, these residents moved from a location where resources for many activities were available into areas that offered few options other than to sit and observe.

Low Mental Status Residents

Prior to the move, low mental status residents spent most of their time out of their bedrooms. Following the move, their use of bedrooms remained stable, but their use of other areas changed. They maintained their proximity to staff by shifting away from frequent use of lounges and dining rooms (down from 48% to 19%) to increased use of the hallways (up from 14% to 38%). Because the new building provided a patio that was fully enclosed and could be entered or viewed from each of the units, low mental status residents increased somewhat the amount of time they spent outside.

In the old building, the residents who are less alert spent 60% of their time asleep or idle, a figure that corresponds closely with other observations of intellectually impaired residents (Brent, Brent, & Mauksch, 1984; Cohen-Mansfield, Marx, & Werner, 1992). Interactions with other residents were infrequent, occurring on the average once an hour. Independent activity was equally rare. Other than the decrease in television viewing noted for all subgroups of residents and an increase in locomotion, low mental status residents showed a stable pattern of behavior. For both television viewing and locomotion, the magnitude of the change was somewhat less for them than for high mental status residents.

Design Complexity and the Use of Space

In general, the increased complexity of the building and distance to social areas seemed to draw high mental status residents from their bedrooms, and this shift in location was accompanied by both increased interaction with staff and increased passive behavior. For low mental status residents, these same features appear to have contributed to their reduced use of designated social spaces. Faced with the challenge of a complex building design, these residents maintained proximity to their rooms and to staff work areas.

EVALUATING THE IMPACT OF PROGRAMMATIC CHANGE

These findings highlight the utility of examining the impact of relocation and other programmatic changes, not only for residents as a group but for subgroups who may have different responses to a given intervention. We have shown how comparisons of building use across different subgroups can point to barriers and inequities that can be the focus of intervention efforts. The similarities of building use between ambulatory and wheelchair-mobile residents suggest that the new building has removed many physical barriers to mobility. Other barriers remained, however. For immobile residents, not enough staff resources were devoted to transporting these residents to social spaces and supervising them there; for low mental status residents, the barriers were building complexity and distance of social spaces from staff areas.

Just as improved physical access can facilitate impaired residents' mobility, advance preparation, as was done for this move, can moderate the stress of relocation and improve residents' adaptation. Hunt and Pastalan (1987) describe how actual and simulated visits to a new facility can help residents develop a cognitive map or mental image of the facility and thus increase residents' feelings of control, adjustment, and even postrelocation survival rates (see also Grant, Skinkle, & Lipps, 1992).

According to Nirenberg (1983), a behavioral skills training program that includes prerelocation visits and specific suggestions about how to perform unfamiliar tasks in a new facility may help impaired residents adjust better when they move into the facility. Most important, in-service programs can make staff more sensitive to the potential negative effects of program change and teach them how to play an ongoing constructive role in facilitating residents' coping behaviors. Pruchno and Resch (1988, 1989) have shown that staff need to attend to the impact of room changes within a facility as well as to relocations between facilities.

We have used a model of person–environment congruence to structure our evaluation of an intrainstitutional relocation. We included important elements of this model by measuring individual competence, aspects of the environment, and resulting behavior patterns. Our findings suggest that a move may have differential impact on residents of varying physical and mental competence. Specifically, mobile and mentally able residents display more diversity of behavior and are better able to accommodate increased environmental demands. Conversely, an increase in environmental resources, such as physical aids to access, may affect residents with moderate impairment more than those whose functioning is intact or those who are very impaired. In chapter 12, we focus further on the congruence model and highlight several issues to consider in the design of residential settings for impaired older persons.

IV

APPLICATIONS FOR PROGRAM PLANNING AND DESIGN

11

Resident Preferences and Design Guidelines

Design and management decisions are influenced by ideas concerning the potential impact of the setting on its residents and by information about what prospective users desire. We have tried to address planners' concerns by developing a conceptually based method for assessing the residential environment and by using the resulting data to examine the impact of environmental features on the social climate of the facility and on resident and staff functioning. In addition, we have obtained information about older adults' preferences for the physical features and policies of group living facilities.

Information about preferences can be used to give residents a voice in planning facilities, as recommended by a number of researchers and planners (Carpman, Grant, & Simmons, 1986; Hartman, Horovitz, & Herman, 1987; Hiatt, 1987; Moore et al., 1985; Palmer, 1981; Sommer, 1983). This information can facilitate the allocation of resources according to the priorities of users and can help reduce mismatch between residents' needs and facility design. For example, Carp (1987b) showed how choice and placement of kitchen appliances were inconvenient and potentially hazardous for older residents of a congregate apartment. She noted that these problems could have been avoided by consulting potential users during the design process. Similarly, knowing residents' preferences for behavioral standards, choice and control, and privacy may help reduce policy conflicts.

This chapter is coauthored by Penny Brennan.

PREFERENCES FOR PHYSICAL DESIGN AND POLICIES

Physical Design Preferences

A few investigators have considered older adults' preferences for physical design features. Regnier (1987) focused on preferences for activity areas in planned housing and found that older adults living in the community preferred spaces designed for everyday activities over those used for more specialized purposes. For example, they preferred a library and large meeting room over a billiard room or gardening space. In surveys of current residents of planned housing, residents emphasized safety, adequate space, and convenience and comfort (Carp, 1987b; Christensen & Cranz, 1987; Cranz, 1987; Nasar & Farokhpay, 1985).

Willcocks and her colleagues (1987) argue that older people's preferences reflect a desire for housing characteristics that are "normal, unexceptional, and non-institutional" (p. 135), that is, features typically available in community housing. They found that residents of residential care facilities strongly endorsed physical amenities, such as windows and doors that can be opened easily, and features promoting social interaction, such as small lounges and small dining room tables. Residents did not want shared bedrooms, movable bedroom furniture, or night lights.

Policy and Service Preferences

Two important policy issues in group living situations are the extent to which behavioral requirements are imposed on residents and the balance between individual freedom and institutional order and stability (chapter 4). Although policies that promote personal control seem to benefit facility residents (chapters 7, 8, and 9), very little is known about residents' preferences in these areas other than their strong desire for privacy (Carp, 1987b; Nasar & Farokhpay, 1985).

More often than not, the initial design and subsequent adjustment of service packages are based on planners' and managers' perceptions of older people's needs and preferences, not on needs-assessment surveys (Avant & Dressel, 1980). Over time, residential facilities usually increase their level of supportive services in response to aging-related decline in residents' functional capacity (Lawton, Moss, & Grimes, 1985; Merrill & Hunt, 1990). When asked directly, older people most frequently endorse health care services, followed by services that support activities of daily living, and then by social and recreational activities (Avant & Dressel, 1980; Lawton, 1975; Regnier, 1987). Economic constraints may increasingly dictate choices among services; systematic information about preferences can help planners, managers, and residents make these difficult choices.

Selecting Individuals to Survey

It is one thing to advocate that facility users have a voice in facility planning; it is another to implement this idea effectively. To begin with, which groups of users should be surveyed? How much do individuals and groups differ in their priorities and preferences? How can the views of residents, staff, and experts be reconciled in the final design and policy guidelines in cases where their opinions diverge?

Current residents of planned housing may seem to be ideal participants in such a survey because of their firsthand experience of a group living situation. But the views of this select group of older people may differ from those of future or potential residents; for new housing to function well over the lifetime of its financing, it must meet the needs of people who are currently middle-aged and whose life experiences and expectations may be quite different from those of older people now in congregate housing (Gelwicks & Dwight, 1982). In addition, the preferences of current residents may be conditioned by what they actually have available, thus perpetuating the status quo. Staff also have firsthand experience with how settings function but may describe their preferences in terms of the work environment rather than the residential environment. Experts in gerontology have broad exposure to a range of settings and to the findings of evaluation research; their views may represent a distillation of the knowledge in this area.

MEASURING DESIGN AND POLICY PREFERENCES

After developing the MEAP, we constructed parallel methods to assess individual preferences for physical design features (chapter 3) and for policies and services (chapter 4). These two preference measures, the Physical and Architectural Features Checklist—Ideal Form (PAF-I) and the Policy and Program Information Form—Ideal Form (POLIF-I), are described here.

We see several ways to use these procedures. They can help planners to set priorities among various environmental resources and to establish policies congruent with resident and staff preferences. They can be used to compare the preferences of different groups, such as residents and staff, or current residents and older people residing in their own homes, or experts in the field. Individual variations in preferences for design and program features can be described. Finally, person–environment congruence can be measured by contrasting the preferences of current or prospective residents with the actual features of a setting; in turn, the degree of congruence can be related to outcomes to determine whether higher levels of congruence lead to better outcomes.

Development of the Preference Measures

We formulated individual PAF-I items by rewording items from the PAF so they would elicit information about respondents' preferences. For example, we rephrased the PAF item "Are there handrails in the halls?" as "Should there be handrails in the halls?" In the same way, we adapted POLIF-I items from the POLIF. For instance, the POLIF item "Is there a residents' council?" became "Should there be a residents' council?" We tried to keep the items as simple as possible and, to facilitate administration of the measures in either a self-report or an interview format, we worded all items as questions.

For both preference measures, we tailored the directions to fit the particular group of respondents. Accordingly, residents and staff of group living facilities were asked to describe the physical features, policies, and services that would be ideal for residents of their facility. For community residents, we outlined the reasons older adults seek group living arrangements (for example, death of spouse, limited finances, or poor health) and asked these respondents to describe the features that would be best for them if they were to seek a group residence. We asked experts to describe a group residential setting that would be best for an older person who might have a physical disability or health problem, but who was capable of independent living.

Older Adult, Staff, and Expert Respondents

We obtained information on preferences from several groups of respondents, including residents of congregate apartment, residential care, and nursing home facilities; congregate apartment residents who rated their preferences for nursing home features; and older adults living in the community. We also included staff working in congregate apartment, residential care, and nursing home facilities, and experts in gerontology.

Here, we focus primarily on three of these groups: older adults living in congregate apartments ($N = 229$), describing preferences for their own facility; older adults living in the community ($N = 205$), rating their preferences for a group residence; and experts ($N = 44$), describing the ideal group residence. We chose this focus because members of the congregate apartment and community groups are best able to understand and respond to the PAF-I and POLIF-I. However, we also report some results for residents of residential care facilities ($N = 153$), residents of nursing homes ($N = 40$), and staff who work in congregate settings ($N = 98$).

Most of the congregate apartment and community residents are in their seventies and eighties (average age = 74 years), 62% are women, and 49% are widowed. As expected, residents of the congregate apartments are older, more likely to be women, and more likely to be widowed than their community counterparts. Both groups of residents have achieved high educational and occupational status; over 60% of them have some education beyond high school, and about 40% reported managerial or professional status. Only 2% are nonwhite (Brennan, Moos, & Lemke, 1988, 1989).

We selected experts by surveying the membership list of the Gerontological Society of America and identifying individuals who had published work in the area of environment and aging.

PREFERENCES FOR PHYSICAL FEATURES

Physical and Architectural Features Checklist—Ideal Form

To construct the PAF-I subscales shown in Table 11.1, we grouped items into the same dimensions as in the PAF. Parallel to the PAF dimensions, the seven PAF-I subscales assess preferences for proximity to community services, for features that increase comfort and active engagement, for supportive features, and for staff facilities. Because we could not develop easily understood items to parallel the PAF space-availability dimension, the PAF-I does not assess preferences in this area. Psychometric information on the PAF-I dimensions is provided elsewhere (Brennan et al., 1988).

Respondents answer each PAF-I item by using a four-point scale that ranges from *not important* to *desirable, very important,* and *essential.* Scores on each PAF-I dimension are the percentage of items that a re-

Table 11.1 Physical and Architectural Features Checklist—Ideal Form (PAF-I) subscale descriptions

Subscale	Description
DEGREE OF PHYSICAL INTEGRATION	
1. Community Accessibility	Preferences for proximity and convenient access to the community and its services
PHYSICAL FEATURES TO IMPROVE CONVENIENCE AND COMFORT	
2. Physical Amenities	Preferences for physical featurese that add convenience, attractiveness, and comfort
3. Social–Recreational Aids	Preferences for physical features that foster social interaction and recreational activities
PHYSICAL FEATURES THAT AID RESIDENTS	
4. Prosthetic Aids	Preferences for the provision of a barrier-free environment and aids to physical independence and mobility
5. Orientational Aids	Preferences for features that help orient residents
6. Safety Features	Preferences for physical features that permit monitoring of communal areas and help prevent accidents
SPACE FOR STAFF FUNCTIONS	
7. Staff Facilities	Preferences for facilities that aid the staff and make it more pleasant for them to maintain and manage the setting

spondent rates as very important or essential. For example, the Community Accessibility subscale is composed of 16 items. Using dichotomous scoring, an individual's score can vary from 0 items (0%) to 16 items (100%) rated as very important or essential. An individual who answers 8 of the 16 items in the scored direction (that is, who believes that 8 of the items are very important or essential) obtains a score of 50%. The dichotomous scoring and computation of percentage scores make the PAF-I subscale scores parallel to the PAF subscale scores.

Congregate Apartment and Community Residents

Figure 11.1 compares congregate apartment and community residents' average scores on the PAF-I subscales. Using these average ratings, we can rank the seven PAF-I dimensions in order of their relative importance to both groups of residents. Safety features and prosthetic aids are rated highest overall, followed by community accessibility, orientational aids,

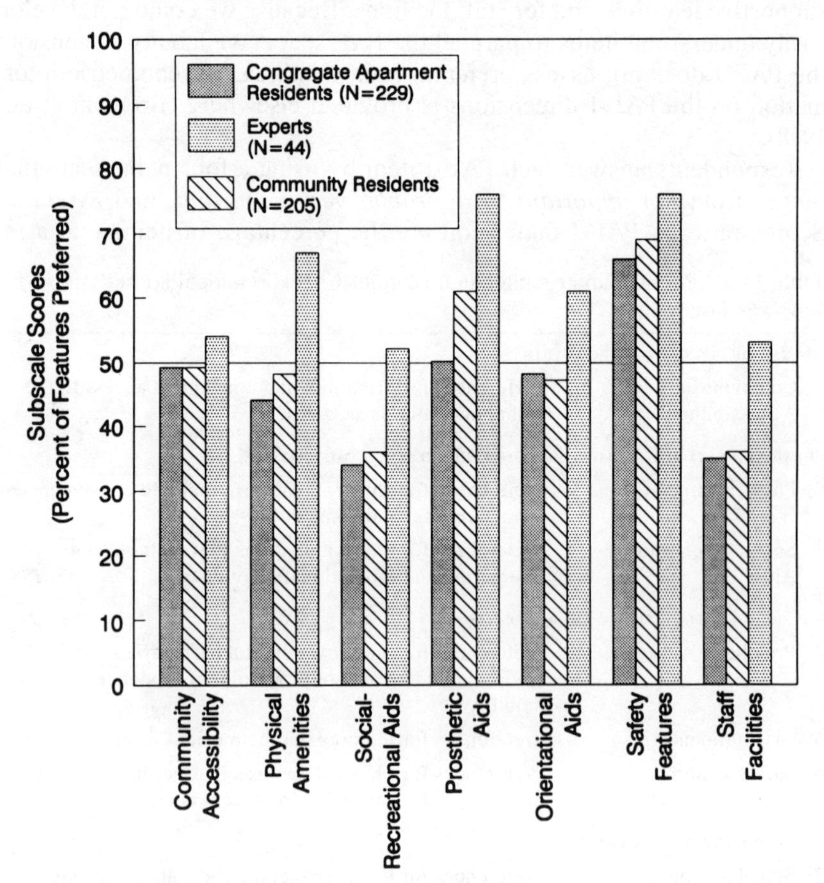

Figure 11.1 Congregate apartment residents', community residents', and experts' preferences for physical features.

and physical amenities. Social–recreational aids and staff facilities are rated somewhat lower overall. Older people in the community closely agree with congregate apartment residents in their preferences, both in terms of the relative priority given the subscales and the strength of their preferences for specific features (r = .88 between ratings of congregate apartment residents and community residents), although community residents have stronger preferences for prosthetic aids.

Supportive Physical Features

The two groups strongly endorse supportive features. Most (70% to 95%) endorse individual safety features, such as well-lighted steps and entries, nonskid surfaces on steps and in bathrooms, smoke detectors, and call buttons in living areas and bathrooms. A majority (50% to 85%) also want such prosthetic aids as handrails in the hallways and bathrooms, lift bars next to the toilet, wheelchair access, and reserved parking for handicapped people; a smaller but still substantial proportion of older adults want to have the front door open and close automatically and to have such bathroom features as a flexible shower head and shower seats. More than 40% rate very important such orientational aids as color-coded floors or corridors, posted instructions for getting into the facility, a large clock in the entrance area, a bulletin board, and a map showing community resources.

Why do the community residents place more emphasis on prosthetic aids than do current congregate apartment residents? Older adults living in the community appear to consider wheelchair accessibility a normative benefit of planned housing. In contrast, congregate apartment residents may interpret such features as accommodations for aging-related decline and, consequently, as unnecessary to the population they think should be served in congregate apartments. They may be less willing to accept prosthetic aids because they want to exclude frail older people from their facilities and preserve their view of themselves as healthy and active.

Features That Promote Active Engagement

Both groups of older people give high ratings to easy access to community resources. In particular, more than 80% rate easy access to public transportation as very important, and more than 40% endorse location near a grocery store, drugstore, and bank. A site near medical services is rated moderately important, whereas proximity to a senior center, park, or movie theater is important to fewer than one-third of the residents. Thus, the priorities for access to community services parallel those for internal facility resources, with health and life maintenance resources rated highest and recreational resources viewed as less important.

Although social–recreational aids are rated as somewhat less important than supportive features or community accessibility, specific aids are strongly valued by older people. In fact, more than 50% of the residents in congregate apartments emphasize the importance of seating in the lobby, a comfortably furnished lounge near the entrance, a patio or open courtyard, small tables around which several people can sit and talk or play games, and a library. Underlining their desire to maintain contact with the outside community, about two-thirds of both resident groups rate adequate parking for visitors very important. However, specific specialized social–recreational features, such as a gardening area, music room, or pool table, are much less likely to be seen as important.

Residents in Nursing Home and Residential Care Facilities

In general, residents in nursing homes and residential care facilities (not shown in Figure 11.1) express preferences that are similar to those of congregate apartment and community residents. However, nursing home residents view prosthetic aids and staff facilities as more important than do residents in residential care facilities, who in turn value them more highly than do congregate apartment residents. These preferences reflect respondents' self-appraised need for a physically supportive environment and more staff. In addition, nursing facilities generally do provide more supportive physical design features and more staff facilities; accordingly, nursing home residents' preferences may be influenced by what is actually made available to them. The overall similarity of preferences among the groups suggests that information from any of them can be used to help planners design and modify the physical features of group living facilities.

Staff Members

Staff members' preferences generally are similar to those of the residents. For example, staff endorse about the same number of prosthetic and orientational aids and safety features as do residents; moreover, consistent with residents' preferences, nursing home staff endorse more prosthetic aids and staff facilities than do staff in residential care and apartments. Despite the differences between residents and staff in such factors as age and social background, staff members mirror fairly closely the views of current residents concerning the importance of these features. Their daily contact with residents and their common experience of currently available conditions in group residences may contribute to these shared views.

Staff members place more emphasis than do residents, however, on features that add to the comfort of the setting, including physical amenities, social–recreational aids, and staff facilities. Staff members have a greater personal stake in the availability of staff facilities, and, not surprisingly, value them more highly. In addition, residents may view amenities and recreational aids as luxuries that they can get along without, whereas staff appear to see them as important contributors to the quality of life in a congregate setting. For example, because of the value staff place on continuing activity in old age, they may favor physical designs that promote social interaction somewhat more than do residents. Consistent with this possibility, Duffy and his colleagues (1986) found that nursing home administrators wanted lounge and dining room layouts to facilitate social interaction; in contrast, residents preferred arrangements that gave them more privacy.

Experts

Planners typically draw upon gerontological research and their own experience as the basis for programming optimal residential settings for older adults (Lawton, Altman, & Wohlwill, 1984). However, preferences of users and experts may diverge in important respects (Avant & Dressel, 1980; Duffy et al., 1986; Gelwicks & Dwight, 1982; Kayser-Jones, 1989; Palmer, 1981; Sommer, 1983; Willcocks et al., 1987). The reliance on expert opinion and failure to consult users can sometimes be to the detriment of older residents (e.g., Carp, 1987b).

Experts and residents generally agree not only on the relative priority of the PAF-I subscales (Figure 11.1) but also on many of the specific items (r = .73 between experts' and apartment residents' ratings of individual features). Experts' preferences are close to older adults' in the areas of community accessibility and safety features—both in terms of subscale scores and average ratings of specific items. For example, like most older adults, the majority of experts (80% to 100%) emphasize the importance of such safety features as well-lighted steps and nonskid surfaces.

On the remaining dimensions, experts give higher ratings than do the older adults. Nevertheless, the relative priority given to specific amenities, social–recreational aids, and orientational aids is very similar for these respondent groups. For example, regarding social–recreational aids, 64% of apartment residents and 80% of experts want seating in the lobby; even though their endorsement rates vary, both groups view this feature as among the most important social–recreational aids.

Regarding prosthetic aids and staff facilities, experts again have higher subscale scores than do apartment residents, but here experts and residents have different priorities within the dimensions. Experts give higher ratings than do residents to wheelchair access and to bathroom features to accommodate wheelchair-bound and frail residents, such as a seat in the shower stall, a flexible shower head, lift bars next to toilets, and turning radius for wheelchairs. In contrast, residents give relatively higher ratings than do experts to handrails in the hallways and automatic front doors.

Several factors may help account for differences in the strength of preferences between experts and the older people currently in congregate apartments. First, as noted in the contrast between congregate apartment and community residents, the apartment residents view wheelchair accessibility as less important than do other groups. This evaluation by apartment residents appears to reflect their wish to exclude nonambulatory residents from their facilities; only 7% feel that nonambulatory residents should be allowed in congregate apartment facilities. In contrast, the experts reflect the view that such congregate settings, as other new construction, should be accessible throughout to handicapped persons.

In addition, experts apparently have achieved consensus about the importance of a number of supportive features, such as handrails and call buttons. Although important to many older people, these features do not rate the nearly unanimous endorsement among residents. Experts' preferences also reflect an image of a facility designed to accommodate a diverse group of older people. Experts thus are more likely than older adults to rate a feature as very important even if it is useful to only a few residents. For example, only 23% of current residents think that garden space is important, but a majority of experts favor including this feature to meet the preferences of that minority to whom it is very important.

The discrepancies between experts and older adults also may stem in part from older people's reluctance to criticize or offer alternatives to existing housing conditions (Carp, 1987a). Older respondents may limit their preferences to existing features of their current residences or to slight improvements on them. In contrast, experts may have exposure to an array of well-designed, resource-rich settings as a backdrop for their preferences.

PREFERENCES FOR POLICIES AND SERVICES

Policy and Program Information Form—Ideal Form

We constructed the POLIF-I subscales (Table 11.2) by grouping items into the POLIF dimensions. Parallel to the POLIF dimensions, the nine POLIF-I subscales assess respondents' preferences about behavioral requirements that might be imposed on facility residents, about individual freedom and institutional order and stability, and about the availability of services and activities. Psychometric information on the POLIF-I dimensions is provided elsewhere (Brennan et al., 1989).

In pilot testing, we found that some respondents strongly opposed particular policies in the areas of expectations for functioning, acceptance of problem behavior, policy choice, and resident control. Accordingly, we phrased response options for items in these four subscales so respondents can rate their preference for these items on a four-point scale that includes *definitely not, preferably not, preferably yes,* and *definitely yes*. On the other POLIF-I items, respondents answer with *not important, desirable, very important,* or *essential*. These different response scales preclude direct comparisons of the two sets of dimensions.

Scores on each POLIF-I dimension are the percentage of subscale items endorsed by a respondent, that is, items rated *preferably* or *definitely yes* for the first four dimensions and *very important* or *essential* for

Table 11.2 Policy and Program Information Form—Ideal Form (POLIF-I) subscale descriptions

BEHAVIORAL REQUIREMENTS FOR RESIDENTS	
1. Expectations for Functioning	Preferences concerning the minimum capacity to perform daily functions that should be acceptable in the facility
2. Acceptance of Problem Behavior	Preferences regarding the extent to which aggressive, defiant, destructive, or eccentric behavior should be accepted
INDIVIDUAL FREEDOM AND INSTITUTIONAL ORDER	
3. Policy Choice	Preferences for options that allow residents to individualize their routines
4. Resident Control	Preferences for level of resident involvement in facility administration and influence on facility policies
5. Policy Clarity	Preferences for formal institutional mechanisms that contribute to clear definition of expected behavior and open communication of ideas
6. Provision for Privacy	Preferences for resident privacy
PROVISION OF SERVICES AND ACTIVITIES	
7. Availability of Health Services	Preferences for health services in the facility
8. Availability of Daily Living Assistance	Preferences for services provided by the facility that assist residents in tasks of daily living
9. Availability of Social–Recreational Activities	Preferences for organized activities within the facility

the last five dimensions. For example, Availability of Health Services is composed of eight items. An individual who answers four of the eight items in the scored direction (that is, who believes that four of the health services are very important or essential) obtains a score of 50%. The dichotomous scoring makes the POLIF-I subscale scores parallel to the POLIF's.

Congregate Apartment and Community Residents

Figure 11.2 compares congregate apartment and community residents' policy and service preferences. Almost all of the respondents believe that fellow residents should maintain relatively independent functioning and conform to high behavioral standards (high expectations for functioning and low acceptance of problem behavior). Most older people want to regulate their own daily activities, and a majority of respondents favor some opportunities for resident involvement and control over decisions. They view privacy as very important, and about half endorse various mechanisms for clear communication. Consistent with their high level of func-

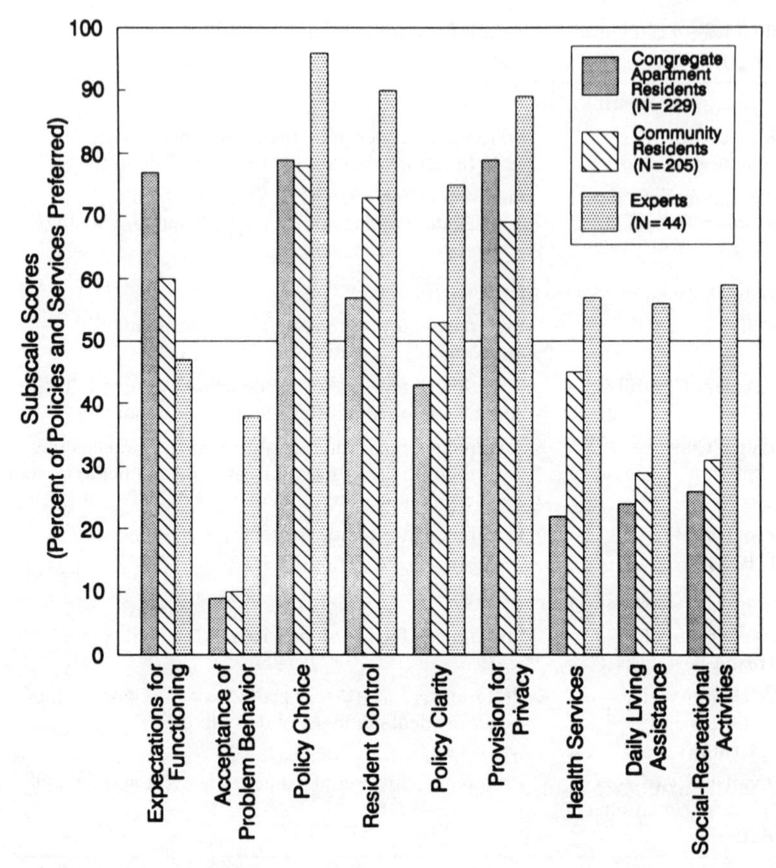

Figure 11.2 Congregate apartment residents', community residents', and experts' preferences for policies and services.

tioning and opinions about the level of disability that should be accepted, respondents rate health services, assistance with daily living, and organized social–recreational activities as less important.

Behavioral Requirements for Residents

Both groups of older adults want high standards for resident functioning and conformity to behavioral norms; less than 10% of the respondents think that behaviors such as intoxication, wandering, and verbal threats should be allowed in group living facilities. However, community residents are willing to permit facilities to set somewhat lower expectations for resident functioning (Figure 11.2). For example, community residents are more likely than congregate apartment residents to view inability to bathe or to feed oneself as tolerable and to accept residents who are not ambulatory. The views of community residents concerning disability are consistent with their stated preference for more supportive physical features.

Individual Freedom and Institutional Order

Both groups of older adults prefer high levels of policy choice (Figure 11.2). For example, a clear majority (60% to 90%) of both groups think that residents should be allowed to wake up when they want to, skip breakfast to sleep late, have some choice of when to eat dinner, sit where they want at meals, and take a bath or shower when they want. Opinions are more divided on alcohol use. Although 59% of the apartment residents think residents should be able to drink alcohol in their own rooms, 41% feel they should not. Regarding alcohol at meals, 53% believe it should be tolerated, but 47% believe it should not be. More of the community residents believe that alcohol use should not be restricted in these ways.

Congregate apartment residents prefer moderate levels of resident control, whereas community residents desire somewhat more. Most older adults (about 85% to 95%) want to have house meetings, committees that include residents as members, and a residents' council that meets regularly. However, a majority of respondents, especially congregate apartment residents, feel that residents should not be involved in selecting new residents, deciding whether to move a resident from one room to another, or selecting staff.

Provision of Services and Activities

Both groups rate relatively few supportive services as very important or essential, but community residents are more likely to endorse health services. Of the health services, respondents feel most strongly about having a doctor on call, having regularly scheduled nurses' hours, and being able to obtain assistance in the use of prescribed medications (endorsed by 28% to 68%). A substantial minority of respondents (17% to 31%) want psychotherapy or counseling, assistance with housekeeping, and legal advice. A majority also wish to have a transportation service linking them to a variety of resources in the community. A moderate proportion of older respondents (20% to 50%) want organized activities such as bingo games, physical fitness activities, classes or lectures, and religious services.

Residents in Nursing Home and Residential Care Facilities

The preferences of nursing home and residential care residents (not shown in Figure 11.2) differ somewhat from those of congregate apartment residents. Consistent with their lower functional ability, nursing home residents endorse greater acceptance of impaired functioning and problem behavior than do residents of apartment and residential care facilities. Residents in nursing homes also rate privacy and choice as less important than do those in residential care facilities; they, in turn, view these policies as less important than do residents in congregate apartments.

To meet their need for more structure and services, nursing home residents are much more likely than apartment residents to endorse the importance of health and daily living services and organized social activities. With the exception of health services, those living in residential care facilities are like apartment residents in their ratings of services.

These findings support the idea that residents prefer environmental conditions that are congruent with their competencies. Thus, for example, nursing home residents prefer more supportive services than do residents of residential care and congregate apartment facilities. Similarly, residents who are functioning better prefer somewhat higher expectations for residents' behavior and somewhat fewer daily living assistance services (Brennan et al., 1988). Thus, a decline in the average functional ability of a resident group is likely to be associated with a shift in preferences toward more daily living services and more lenient standards regarding physical impairment and problem behavior.

In general, community residents' preferences for policies and services more closely correspond to residential care and nursing home residents' preferences than to those of congregate apartment residents. Congregate apartment residents apparently want their facilities to resemble independent community living as much as possible and to avoid policies and services that increase the institutional quality (cf. Willcocks et al., 1987). In contrast, community respondents see a greater gap between community and congregate living. They seem to feel that to justify the move into a congregate setting, the facility should tolerate some disability, provide supportive health services, and have institutional mechanisms for giving residents control and for communicating clearly.

Staff Members

In the area of policies, staff members' preferences closely resemble those of residents in similar types of facility. For example, residents and staff have similar views of the desirable levels of expectations for functioning, acceptance of problem behavior, policy choice, and provision for privacy. The similarity of their views, which may reflect their daily contact and shared environment, enables staff to provide a valuable perspective for planners of group residences.

In the area of services, however, staff members and residents differ in their preferences. Consistent with the staff's service orientation, staff members are more likely than residents to endorse the provision of daily living assistance services and facility-planned social–recreational activities. On average, staff rate 65% of the planned social activities as very important or essential; in contrast, only 32% of the residents give such high ratings to these activities. The staff in nursing homes also view health services as more important than do the residents (74% vs. 59%).

Experts

Experts differ from apartment and community residents on all nine of the POLIF-I dimensions (Figure 11.2). Compared with congregate apartment and community residents, experts prefer facilities with fewer expectations for resident functioning and higher acceptance of problem behavior. Experts are also more likely than these older adults to endorse choice in daily routines, resident influence over facility policies, formal mechanisms for communication, and provisions for privacy. In addition, the experts give a stronger endorsement than do the older adults to offering more health services, assistance in activities of daily living, and organized activities. As with the PAF items, however, experts' relative priorities are similar to those of apartment residents (average $r = .68$ for the nine subscales); they differ most in the area of social–recreational activities.

The differences between experts and older adults in their preferred policies and services parallel their respective preferences for physical design features. Older adults may have difficulty acknowledging increased impairment and dependency and therefore prefer an environment with few accommodations to residents with functional impairment. In contrast, experts acknowledge that some older adults need services and that these services may be provided most efficiently in a congregate setting. In fact, experts may find it difficult to justify specialized residences for older people unless supportive services are available and functional impairment is accepted.

Experts endorse a highly supportive setting (in terms of expectations, physical features, and services) and also want a facility with a high level of challenge (choice in daily routines and control of facility operation). Such a setting could serve a diverse group of older people by offering sufficient challenge to the most intact and supportive services to all but the most disabled. Such an accommodative housing model would allow residents to maintain long-term residence despite declines in functioning that may accompany aging (Lawton et al., 1980). In contrast, most older people appear to prefer a facility tailored to a narrower range of needs and abilities, which is consistent with the model that has generally guided planning and development of facilities.

CONGRUENCE BETWEEN ACTUAL AND PREFERRED FACILITIES

One reason that we developed the PAF-I and POLIF-I was to compare residents' preferences with the actual features of settings. Such comparisons can identify ways to increase the congruence between residents' preferences and the physical features, policies, and services in group living facilities. To provide two examples, we first compare the preferences of congregate apartment residents with the characteristics of Amity

House, then compare the preferences of residents of residential care facilities with the characteristics of Maude's Community Care Home; we described both of these facilities in chapters 3, 4, and 5. (For comparisons between the preferences of an individual resident and the characteristics of a congregate apartment, see Moos & Lemke, 1992a, 1992c.)

Amity House and Residents' Preferences

Amity House residents are similar in background to the congregate apartment residents who completed the PAF-I and POLIF-I, but the Amity House residents are functionally more intact and more active in the community.

Physical Features

As Figure 11.3 shows, Amity House generally has the physical features that congregate apartment residents think are important. Although its rural location makes it more isolated from community resources than

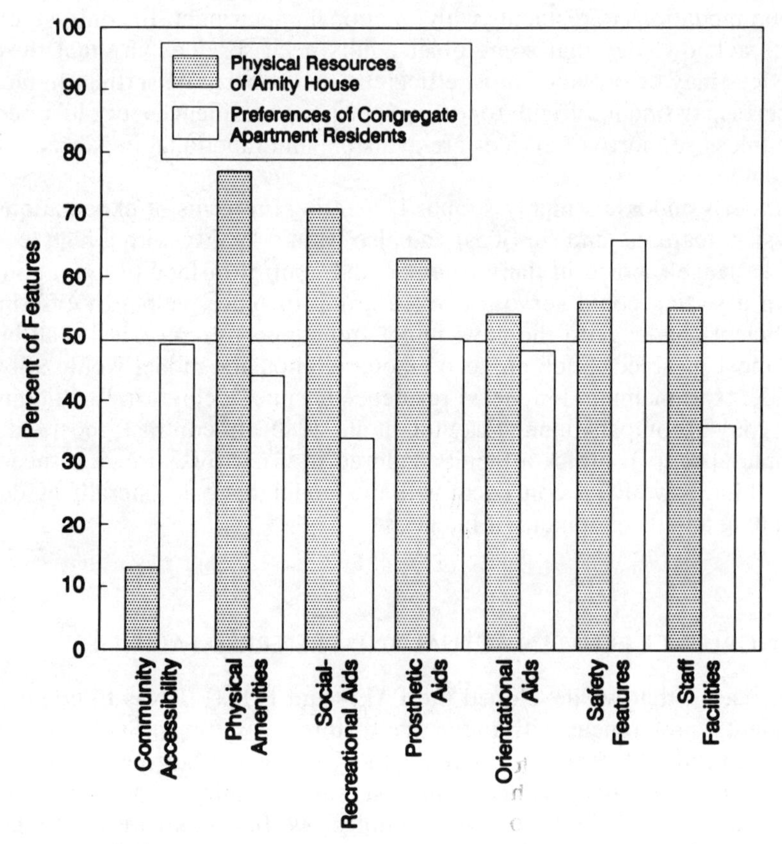

Figure 11.3 Physical resources of Amity House and preferences of congregate apartment residents.

these older people would like (13% of resources easily accessible vs. a preference for about 50%), Amity House provides ample amenities and social–recreational aids. For example, Amity House has 71% of the social–recreational aids included on the PAF; only 34% of these are rated very important by these respondents. Respondents rate slightly more safety features very important (66%) than are offered by the facility (56%); in particular, a majority feel that the entry and outside seating should be visible from a staff area, that hallways should have smoke detectors, and that bathrooms should have call buttons, but Amity House does not provide these.

Policies and Services

There are a number of discrepancies between the policies found in Amity House and the preferences of congregate apartment residents (Figure 11.4). The residents prefer a much lower level of acceptance of problem behavior than is shown; whereas Amity House's policy is to accept a wide

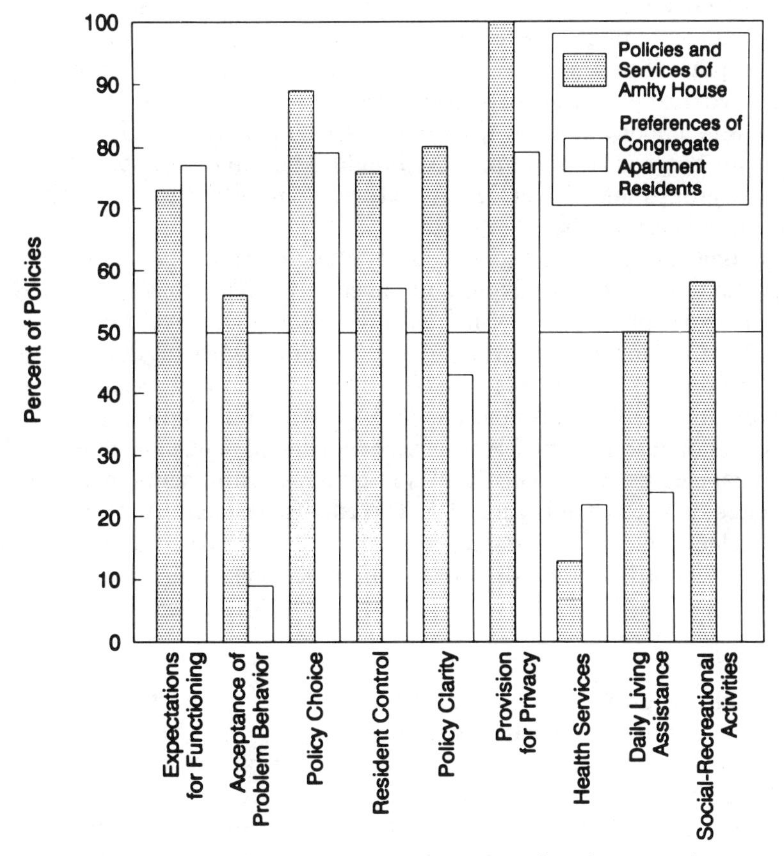

Figure 11.4 Policies and services of Amity House and preferences of congregate apartment residents.

variety of behavior, including wandering around the building or grounds at night, creating a disturbance, or refusing to bathe regularly, a majority of apartment residents prefer that such conduct not be allowed. Amity House offers somewhat more flexibility in daily routine than is preferred by apartment respondents and substantially more involvement in decision making. For example, although Amity House involves residents in dealing with residents' complaints, making rules about the use of alcohol, and moving residents within the facility, only about one-third of apartment respondents feel that residents should be involved in these areas of decision making.

Compared with the levels rated very important by apartment residents, Amity House also puts more effort into providing avenues for clear communication and ensuring privacy. Amity House has a handbook for residents, an orientation program for new residents, and a newsletter, but fewer than half the respondents rate these provisions very important.

The services offered by Amity House exceed the level rated as very important in two areas—daily living assistance and organized activities. Amity House offers all forms of daily living assistance that at least 25% of the residents rate as very important, as well as some services that are very important to fewer residents, such as assistance with banking and a daily breakfast plan. Similarly, Amity House provides almost all activities rated very important by at least one-fourth of the residents; the exceptions are bingo games and parties. In addition, Amity House offers discussion groups and a social hour, activities important to a smaller proportion of apartment residents.

To some extent the policies and services at Amity House may be a response to the views of experts involved in planning that setting and to the particular preferences of its resident population. Except on expectations for functioning and availability of health services, Amity House's levels of policies and services are closer to the levels endorsed by experts than by congregate apartment residents. It would be important to assess the views of current Amity House residents and to determine whether the discrepancies between their ideal and actual environments match those we noted in discussing Figure 11.4. The current residents may be a somewhat select group; for example, resident control may be more important to them than it is to the normative sample of congregate apartment residents.

Implication for Change

Amity House provides a physical environment that meets the expressed needs of congregate apartment residents, except for the difficulties imposed by its rural location. In its policies, Amity House may pose a challenge to some residents by providing higher-than-preferred levels of resident control; in fact, according to the Sheltered Care Environment Scale—Ideal Form (SCES-I; Moos & Lemke, 1992f), the social climate of Amity House provides more resident influence than residents of congregate apartments and residential care facilities tend to prefer. Overall, the environment provides high levels of support and stimulation aimed at an active and engaged resident population.

In its materials to potential applicants, Amity House could provide information about its unique characteristics in order to facilitate a match between residents' preferences and the current environment. Administrators also need to consider whether they are missing a portion of their target population because of the challenges that Amity House presents, as well as whether the program, as it is currently carried out, is not meeting the needs of some current residents.

Maude's Community Care Home and Resident's Preferences

Maude's Community Care Home, a small, privately run residential care facility, was built in the 1970s. Its residents differ in a number of respects from the respondents who completed the preference questionnaires for residential care facilities. Maude's residents are much older and have more impaired functioning; they also have lower socioeconomic status.

Physical Features

Not surprisingly, given its recent construction and the fact that it was built as a residential care facility, Maude's provides the physical features that are important to respondents currently living in residential care facilities, as shown in Figure 11.5. In fact, it exceeds their expressed needs in a number of areas, including its accessibility to community resources, the availability of social–recreational aids, and safety features. Thus, the building reflects effective allocation of resources to satisfy the wishes of individuals who are likely to be residents.

Policies and Services

As shown in Figure 11.6, there are large discrepancies between residential care facility residents' preferences and Maude's policies and services. Respondents prefer higher behavioral standards than are imposed at Maude's. For example, a majority of residential care facility respondents feel that residents should be expected to be ambulatory, able to dress without assistance, and free of confusion or disorientation. The administrator reports that he would accept residents who are not able to function at this level, although none of the current residents experience such impairments except for one who is mildly confused. Similarly, although acceptance of problem behavior is moderate, respondents prefer less; for example, residents would be allowed to stay at Maude's even if they created a disturbance, refused to take their medications, or wandered at night, but fewer than 15% of respondents believe that such behaviors should be allowed.

In contrast, Maude's offers much less choice and control than respondents would like. Staff members set a time for getting up, going to bed,

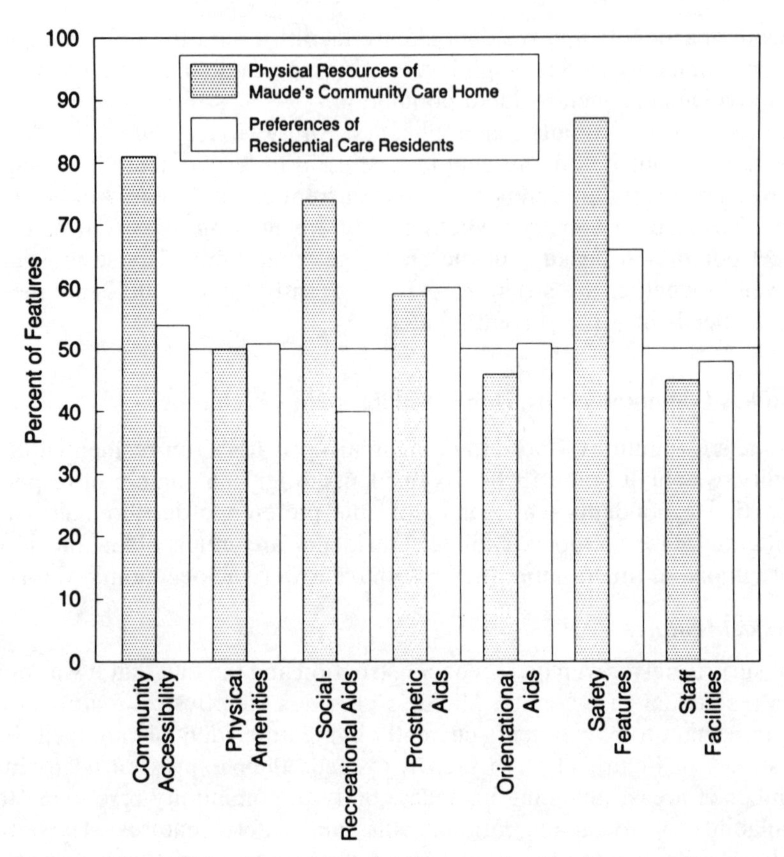

Figure 11.5 Physical resources of Maude's Community Care Home and preferences of residential care residents.

bathing, and eating, and they discourage residents from rearranging their furniture; a majority of respondents feel that they should be offered greater leeway to decide these matters for themselves. A substantial majority want house meetings and a residents' council and feel that residents should be involved in chores, in planning activities, and in dealing with residents' complaints; however, Maude's does not provide these opportunities. Maude's provides the levels of policy clarity and privacy that respondents prefer.

Finally, in the area of services, respondents would like more health services than Maude's offers; for example, more than half the respondents desire regularly scheduled nurse's and physician's hours and physical therapy, services not available at Maude's. Maude's does provide the types of daily living assistance and organized activities that are very important to a majority of respondents, as well as some that are endorsed by fewer respondents. For example, the facility helps residents with personal grooming and handling spending money, services that are rated very important by about 20% of respondents in residential care facilities.

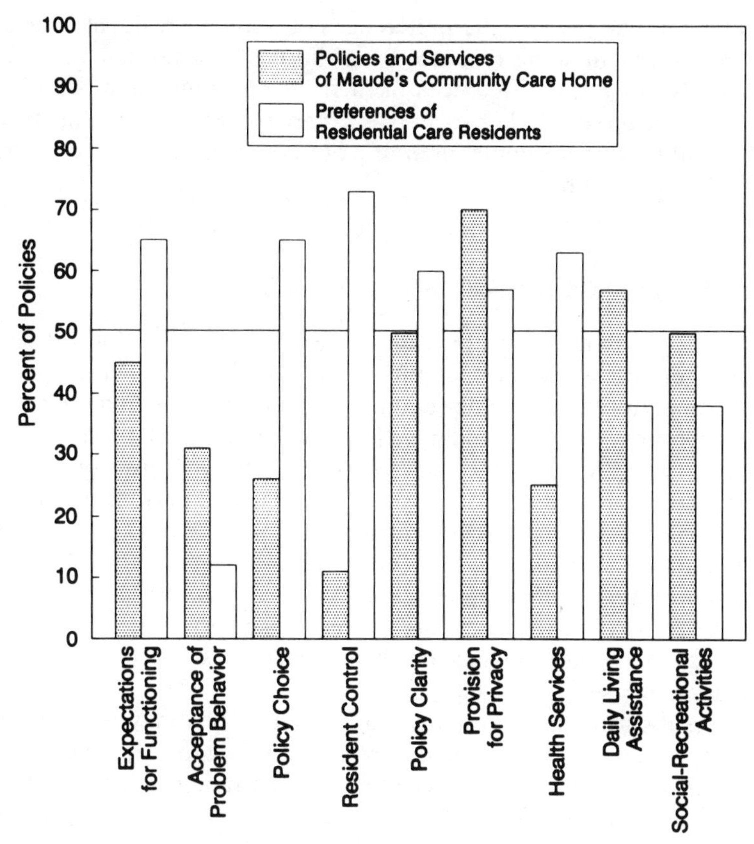

Figure 11.6 Policies and services of Maude's Community Care Home and preferences of residential care residents.

Implications for Change

To identify potentially useful interventions, administrators can consider information from a variety of sources, including these profiles showing areas in which Maude's differs from the preferences of residential care facility residents. In addition, the low level of perceived independence (chapter 5) and information on physical design and policy features related to independence (chapter 7) might help focus intervention efforts. Specifically, Maude's could increase the number of social–recreational aids it provides and modify its policies to allow for flexibility of daily routines and resident involvement in decision making; these changes could be expected to increase the level of resident independence.

We noted that the service priorities for residential care facility respondents are different from the services actually available at Maude's. Without increasing expenditures, the administrator could reallocate resources so that service availability more closely matches the respondents' priorities. However, before undertaking such changes, the administrator should determine the specific preferences of Maude's residents, because they may differ in some respects from those of the normative sample, particularly given the differences in their current functioning and background characteristics. (For examples comparing the social climate of a facility and its own residents' or staff's preferences, see Moos & Lemke, 1992f; Waters, 1980b.)

IMPLICATIONS FOR ASSESSMENT AND THEORY

This work on preferences illustrates several important points. It shows that planners can obtain meaningful information about the design and programming preferences of older adults living in the community and those living in group residential facilities. The findings show that some features are much more important to older adults than others. Thus, for example, safety features and prosthetic aids are rated very important, as are policies that encourage personal control and discourage deviant behavior. Design features such as physical amenities and social–recreational aids and program features such as social–recreational activities are rated as somewhat less important. In general, preferences appear to reflect a hierarchy of needs, with life-maintenance needs ranking highest and needs for social stimulation and self-actualization ranking lower.

The results also show that individuals and groups may differ in their preferences and priorities, in some areas, according to their level of functional impairment and the facility type they are rating. For example, occupants of nursing homes prefer lower expectations for functional independence and more liberal standards for acceptable behavior, more services, and more structured daily routine than do residential care, congregate apartment, and community residents. In contrast, residents in all three facility types have similar preferences for physical design features, with the exceptions of prosthetic features and staff facilities.

Thus, planners should obtain information about policy and service preferences mainly from current residents of facilities similar to those being developed. However, current residents select congregate facilities in part because conditions there meet their needs; focusing only on current residents may tend to preserve the status quo and leaves untapped the preferences of older people whose needs are not met by existing congregate settings. Planners should consider the views of staff and experts working in this field. And to keep abreast of generational changes, planners also need to monitor the preferences of older adults who are likely to become users of supportive housing services.

Ultimately, no single model of congregate housing is likely to meet the expressed needs of all older people. Opinions are divided concerning some features and policies, such as wheelchair accessibility, alcohol use, whether meals should be served, and what medical services should be offered. These features and policy areas pose dilemmas for planners and administrators. What obligations does a facility have to accommodate those whose views are in the minority? How can the costs of design, policy, and service decisions be balanced against the benefits to different subgroups?

In contrast with other respondent groups, experts endorse an accommodative model of congregate housing. Specifically, experts prefer richer physical resources and both more supportive and more challenging social environments than do older adults. Our findings generally show positive outcomes for older people living in such settings; however, to supplement the data on preferences, planners need more information on the impact of these features on various aspects of well-being for subgroups of older people.

Our results suggest that higher-functioning residents may resist implementation of an accommodative housing model, wherein more services are supplied over time to compensate for residents' aging-related deficits. Residents who are somewhat more impaired may accept this approach more readily. As in the education of children with special needs, mainstreaming is likely to be met with resistance from some groups of users. Some older people are intolerant of individuals with disabilities and make heroic attempts to hide their own declines in functioning when they face the prospect of being relocated to more-supportive housing.

Finally, residents' preferences can be compared with the actual features of planned housing. As we illustrated with the profiles of Amity House and Maude's, such comparisons can help identify areas in which to initiate change. Similarly, prospective users can be surveyed in order to develop program changes to ensure continuing appeal to new applicants (e.g., Merrill & Hunt, 1990). In general, evaluations that compare resident preferences to the actual features of a facility can guide interventions aimed at increasing the congruence between people's preferences and facility features. Such congruence may contribute to resident satisfaction (e.g., Conway, Vickers, & French, 1992).

12

Implications for Theory and Design

We have presented a conceptual model to guide the evaluation of residential facilities for older people, developed new measures to characterize these programs, and applied these measures in three types of settings (nursing homes, residential care facilities, and congregate apartments). Throughout our work, we have emphasized the interplay of person and program factors in determining resident outcomes. In this chapter, we use the conceptual framework to review and integrate the major findings of the research program. Then we focus on the implications of the framework and findings for program evaluation and design and conclude with a discussion of some important future directions.

A MODEL OF RESIDENT ADAPTATION

In chapter 1, we pointed out that a growing number of older adults live in supportive residential facilities and suggested that these diverse facilities need to be evaluated within a common framework. Our conceptual framework focuses on the personal and environmental factors that contribute to residents' adjustment and well-being. The framework encompasses objective environmental characteristics and the social climate of residential programs, personal characteristics (such as residents' social resources, functional abilities, and preferences), how residents cope with living in a facility, and residents' outcomes. In addition, we have adopted a general person–environment congruence model; that is, we posit that the more competent individuals are able to construct and adapt to a wide range of environmental conditions, whereas less competent individuals are more constrained by environmental factors and need more supportive resources to function optimally.

CONCEPTUALIZING RESIDENTIAL PROGRAMS

To begin to apply the model in residential facilities, we conceptualized the program environment as composed of four main sets of factors and developed a set of procedures for measuring them. The objective environment includes the aggregate characteristics of the residents and staff, the program's physical design, and its policies and services. These three sets of factors combine to influence the fourth domain, the social climate. We developed the Multiphasic Environmental Assessment Procedure (MEAP) to assess these four domains of variables and applied it to residential facilities for older people, including 262 community and 81 veterans facilities.

Resident and Staff Characteristics

The Resident and Staff Information Form (RESIF) uses information about residents' demographic characteristics to assess the social resources and heterogeneity of the resident population; it also measures residents' aggregate functional abilities and activity involvement and the resources available from the staff (chapter 2). The facility types show considerable overlap in the demographic characteristics of their resident groups. As expected, however, residents in congregate apartments function more independently and engage more in informal activities than do residents in the other types of facilities. Nursing home residents are the most functionally impaired and least likely to initiate leisure activity; residential care facility residents are intermediate in these respects.

To manage their much more impaired resident populations, nursing homes have a higher staff–resident ratio than do residential care facilities, and the staff are more experienced and diverse. Residential care facilities in turn have a higher staffing level and more staff resources than do congregate apartments.

Physical Features

To assess the physical environment, we developed the Physical and Architectural Features Checklist (PAF), which taps the facility's proximity to community resources, its comfort, safety, and space allowances, and the ease of orientation and of use by those with handicaps (chapter 3). On average, the three facility types vary somewhat in their physical features. Nursing homes have the most prosthetic aids and space for staff functions. Residential care facilities have the most space for residents, and congregate apartments are most accessible to community resources.

Policies and Services

The Policy and Program Information Form (POLIF) measures the behavioral requirements for residents of a facility, the balance between individual freedom and institutional order, and the provision of services and activities. The findings show that policies and services differ sharply between facilities that offer different levels of care (chapter 4). Nursing homes have lower expectations for independent resident functioning than do residential care facilities, which have lower expectations than do congregate apartments. Consistent with these differences, the residents in nursing homes have access to more health and daily living services and more planned social activities, but they are given fewer opportunities for choice, control, and privacy. The moderately impaired residents in residential care facilities have somewhat higher levels of privacy and lower levels of services compared with nursing homes. Congregate apartment residents exercise more choice and control and enjoy more privacy but have access to fewer services than residents in the other types of facilities.

Social Climate

In addition to asking observers and administrators about the characteristics of residential facilities, we focused on the perceptions of the residents and staff who live and work in these settings (chapter 5). The Sheltered Care Environment Scale (SCES) characterizes a facility's social climate in terms of the quality of personal relationships (cohesion and conflict), the relative emphasis on aspects of personal growth (independence and self-disclosure), and mechanisms for system maintenance and change (organization, resident influence, and physical comfort).

Compared with the two other types of facilities, the conditions in nursing homes produce more conflict and confusion. However, although their residents are quite impaired, nursing homes establish social environments that are as cohesive as those in congregate apartment and residential care facilities. Congregate apartments, consistent with their residents' better functional abilities, tend to establish social climates with more emphasis on resident independence. Despite the emphasis on resident choice and control, however, apartment residents report no greater influence in the facility than do residents in nursing home and residential care facilities. In general, staff responses show the same pattern of differences between facility types. Compared with residents, staff members tend to report that facilities are more cohesive and have a higher level of conflict, that residents openly discuss their feelings and concerns, and that residents exercise more self-direction.

These results illustrate several important points about facility social climate. First, multiple pathways may lead to a particular social climate. Thus, for example, despite the very different objective conditions in nursing homes and congregate apartments, friendly relationships and a sense of resident influence are equally likely to develop in either setting type. Even in areas where the facility types differ on average, they overlap substantially. The second point is that social climate judgments are relativistic and contextual in nature. For example, congregate apartments and nursing homes do not overlap in the actual functional abilities of their resident populations, but residents' reports of their independence do overlap, which suggests that to some extent nursing home residents judge the facility emphasis on independence relative to their expectations. In the same way, staff members view a setting within a different framework than do residents and, consequently, differ from residents in how they perceive a facility.

Ownership and Size

In addition to differences between facility types, we identified some important differences in resources related to ownership and facility size (chapter 6). Compared with proprietary facilities, nonprofit facilities provide a somewhat more comfortable, supportive, and spacious physical environment. Policies in nonprofit facilities are more accepting of problem behavior and more likely to promote residents' personal control. Nonprofit facilities also provide more health services, a somewhat greater variety of social activities, and more staff resources. As expected from these differences, residents in nonprofit facilities see their facilities as more cohesive and resident-directed and somewhat better organized than do residents in proprietary facilities. Veterans facilities offer more health services but provide less privacy, less policy choice, and a less harmonious environment than do either proprietary or nonprofit facilities.

In general, these findings confirm the advantages of nonprofit facilities, particularly in areas that are more difficult to regulate, such as the comfort of the physical environment, policies that enhance resident self-direction, and the quality of the social environment. They suggest that the profit motive may compromise the quality of care and that the ethos of service to the consumer found in nonprofit organizations may lead to more humane residential facilities. The conditions in nonprofit facilities also illustrate that higher levels of support can coexist with higher levels of autonomy.

Our findings show that large size does not necessarily lead to a more restrictive and less individualized environment. In general, larger community facilities have a more varied array of physical features and services and more staff resources; larger facilities also provide more opportunities for resident self-direction and independence. However, residents and staff in larger community facilities reported more conflict and somewhat poorer organization. Thus, larger facility size may promote resident independence but compromise good organization and cause more strain in relationships.

Selection and Allocation

Selection and allocation processes result in wide variations among congregate residential facilities in the characteristics of their residents and staff and the resources available to them. In general, two processes operate to increase the homogeneity of resident populations within facilities and the correspondence between resident and facility characteristics. One process involves residents' needs; the other involves predisposing and enabling factors.

Need Factors

Need for assistance, as reflected in residents' functional ability, is the strongest determinant of the type of facility individuals enter and, consequently, of the pattern of resource allocation they encounter. The populations of impaired residents in nursing homes have access to a higher staffing level provided by individuals with more training and diversity of background. Nursing homes provide more prosthetic aids and space for staff functions but are more distant from community resources and have less space for resident functions. Nursing homes also provide a high level of formal services to address their residents' health care needs and functional impairments, but they allow residents less privacy and personal choice and involve residents less in decision making.

As facility types are currently structured, trade-offs are made between features that provide support and those that enhance autonomy. Facilities serving more functionally impaired residents tend to emphasize supportive resources (higher staffing, prosthetic aids, services) and the communal environment (facility-organized activities and amenities and social–recreational aids in public areas). Facilities serving more functionally intact residents tend to emphasize autonomy (proximity to community resources, amenities that support independence, and flexibility of routines) and the private, personal environment (amenities in individual rooms).

Two implicit assumptions seem to underlie this pattern of resource allocation. One is a needs-hierarchy model; the other is the idea that there is a necessary trade-off between security and communality on the one hand and autonomy and individuality on the other. According to these views, older people with physical disabilities or significant intellectual impairment have primarily security and life-maintenance needs, and facilities that serve them should be designed to compensate for these disabilities. In doing so, personalization and resident autonomy must inevitably be sacrificed. In contrast, individuals with relatively intact functioning have primarily autonomy and life-enhancement needs; thus, their facilities should be designed to foster autonomy and self-expression and must necessarily minimize supportive features.

This model of resource allocation is manifested not only in the current conditions of congregate facilities but also in the preferences of residents and staff. For example, residents and staff see prosthetic features, services, and staff facilities as more important in nursing homes and policy choice and privacy as more important in apartments (chapter 11). In contrast, experts adhere to an accommodative model in which a range of functioning is allowed and high levels of both support and autonomy are provided.

Within facility types, some resources are allocated on the basis of functional ability or needs–resources congruence, but the pattern is less clear than in the comparisons between facility types. For a given facility type, those that serve more functionally impaired residents provide higher staffing and more prosthetic aids and offer less privacy, and less opportunity for resident control. However, contrary to expectations from a needs-based congruence perspective, within levels of care, residents who are more impaired tend to be in facilities that provide less daily living assistance and no more health services.

A number of factors help explain the weaker influence of needs–resources congruence in determining the allocation of resources within levels of care. Because individuals who decline in functional ability do not necessarily relocate to more service-rich facilities, the aggregate functional ability of facility residents is likely to decline over time. However, the level of facility resources may be less elastic or have a long lead time to implement changes such as hiring and training staff to provide specialized services. In addition, relatives or community agencies, rather than the facility may provide new services, so residents may continue to have their needs met. Another factor that reduces the impact of needs–resources congruence is the counteracting influence of enabling factors in determining resource allocation.

Predisposing and Enabling Factors

The second process, based on predisposing and enabling factors, allows more privileged individuals to live together and to obtain access to better staff resources and physical features, more liberal policies, and more services. These residents probably have more leeway in facility selection, place more emphasis on policies that provide personal control in selecting a facility, have the financial resources to purchase more services, and support these aspects of facility management once they become residents.

This process helps to explain some of the variation in resident characteristics and staffing in facilities that provide the same level of care. Thus, once an appropriate level of care is selected, predisposing factors (such as gender and age) and enabling factors (such as married status and higher educational and occupational status) are as important as need factors in providing residents access to particular facilities and their resources. This process does not appear to entail a trade-off between autonomy and supportive resources.

The Impact of Residential Programs

Objective features help to determine the social climate that develops in a setting (chapter 7); these factors together influence the adjustment and well-being of the residents and staff who live and work in them. We examined some of these influences on resident groups (chapter 8) and on individual residents (chapter 9). We also evaluated an intrainstitutional relocation and described the varied outcomes for residents who differed in mobility and mental status (chapter 10).

The Determinants of Program Climate

Residential programs vary widely in the quality of relationships, the emphasis on resident self-direction, and the level of organization. Chapter 7 focused on how other program factors might contribute to these differences in facility social climate.

Residents' social resources and capabilities influence the social climate that develops in a residential facility. Resident groups that are more socially privileged and that include a higher proportion of women tend to develop social climates that are more harmonious, resident directed, and well organized; high-functioning resident groups also tend to give residents more sense of influence and to be better organized. These resident groups may be better able to take the initiative, respond to staff encouragement of independence, and voice effectively their opinions and concerns. In turn, staff may be more likely to initiate and maintain policies that support resident independence and influence when they see residents as capable of using these opportunities.

More cohesive and resident-directed social climates also tend to emerge in facilities with more staff resources, more space for resident and staff functions, and more physical features that add comfort and provide opportunities for recreation and social interaction. Clearer policies, policies oriented toward personal control, and more planned social activities also contribute to a harmonious and resident-directed social climate. We believe that the social climate is key because it is likely to have an important influence on residents' adjustment and well-being. By developing a better understanding of some of the determinants of social climate, these findings can help residents and staff shape a more beneficial social environment.

Personal Control Policies, Social Climate, and Residents' Adaptation

In group-level analyses, we found that personal control policies (privacy, policy choice, resident control, and policy clarity) were associated with better resident adjustment, greater involvement in activities in the community, less involvement in facility-planned social activities, and less reliance on the health and daily living services provided in the facility. In general, a cohesive and resident-directed social climate was also related to better resident adjustment and less use of facility services (chapter 8).

Residents' aggregate functional abilities moderated some of these relationships. Specifically, policies that provide more personal control and a social climate oriented toward independence were more closely associated with adjustment and use of services among functionally able than among functionally impaired resident groups. In facilities with relatively intact or only moderately impaired residents, policies that emphasize resident control and an emphasis on independence were associated with better adjustment and less reliance on supportive services. In facilities with residents who are more impaired, however, aggregate functional capacity was the main determinant of adjustment and service utilization; personal control policies and an emphasis on independence had relatively little influence on these aspects of adaptation for impaired resident groups. These findings support a needs-hierarchy model and the idea that these particular environmental resources are more salient for functionally competent residents.

Residents' Engagement in Social Activities

We also used the congruence model to examine the personal and facility factors associated with individual differences in residents' involvement in activities (chapter 9). Women residents and better-educated and functionally intact residents were more involved in informal activities in the facility or community. Residents were also more likely to participate in informal activities when other residents were more involved in such activities. In addition, more privacy and a harmonious, resident-directed, and well-organized social climate were associated with greater involvement in informal activities. Resident control and policy clarity also were associated with more involvement in activities in the community.

We also identified a pattern of interaction between the policies and social climate of the facility, the functional abilities of individual residents, and their participation in informal activities. For residents who were functionally more independent, personal control policies and an emphasis on cohesion and independence were positively related to involvement in informal activity; these relationships weakened as functioning declined. Very impaired residents showed little or no change in informal activity associated with increases in these particular program factors, which appear to be more salient for the more intact residents.

The findings were somewhat different for residents' participation in facility-planned activities. One pattern of interactions showed that the more impaired individuals are drawn into facility activities in the presence of environmental resources such as active group participation and an emphasis on resident influence and organization. For example, individual participation in facility activities was higher in better-organized facilities, and this relationship was strongest for impaired residents. As functional ability increased, this relationship weakened. Thus, these aspects of the program appear to address the particular needs of impaired residents and to reduce the barriers that keep them from participating in the formal activity program. These features are less salient for residents whose functioning is intact.

A second pattern of findings implies that impaired residents were less likely to participate in planned activities when facility policies gave them more choice in their daily routines. These results were similar for resident control and privacy; they suggest that for residents who are functionally impaired, personal control policies may reflect insufficient structure and support for full participation, as well as an activity program geared toward the abilities of high-functioning residents. Thus, personal control policies may operate as an environmental demand that reduces impaired residents' involvement in organized activities.

Coping with Relocation

In chapter 10, we used the congruence model to structure an evaluation of an intrainstitutional relocation. The move had differential impacts on residents of varying mobility and mental status. Specifically, an increase in prosthetic aids and fewer physical barriers to access led to an increase in ambulatory and wheelchair-mobile residents' locomotion. However, these environmental changes had no such effect on immobile residents, in part because the new building was more dispersed and the immobile residents did not receive increased assistance from staff to transport them. For mentally able residents, the added complexity of the new building and increased distance between staff areas and social spaces led to increased use of social space; in contrast, mentally impaired residents used these areas less in the new building.

Overall, mobile residents and residents of high mental status displayed more diversity of behavior and were better able to accommodate increased environmental demands. Wheelchair-mobile residents had been restricted by aspects of the old building; when resources were added to help them compensate for their disabilities, they showed changes in their behavior not shown by other groups.

Resident–Facility Congruence

The findings show that residents who are functionally more independent are more likely to be affected by a program's emphasis on individual choice and independence; in contrast, residents who are more functionally impaired are influenced more by the level of support and structure in the program. Accordingly, a demanding and stimulating environment may enhance adaptation among individuals who are functioning relatively well, whereas very impaired individuals may benefit most from a well-organized setting with sufficient resources to help them overcome obstacles to optimal adaptation. In general, residential facilities should try to establish an active, involving environment and foster residents' self-direction; however, it is also important to develop supportive programs in which very impaired individuals can achieve their optimum level of adaptation.

The differential needs of high- and low-functioning individuals have been generally recognized. In response to these differences, facility planners commonly try to define a distinct, homogeneous grouping that a residential program will serve. However, as we have noted, a number of barriers prevents the creation of such homogeneity: Functioning is multifaceted, difficult to assess, and changes over time; and as a practical matter, settings serve mixed populations. We argue instead for recognizing the varied needs of facility residents in program design.

Despite the tension that exists between support and autonomy, facilities may be able to provide high levels of both in order to meet the needs of a heterogeneous population. In fact, nonprofit programs and those serving residents who are more privileged apparently do provide such a mixed model, a finding that suggests that facilities are more likely to make trade-offs not because of residents' needs but because of the program philosophy or limited resources. Our findings also suggest that residents will take advantage of those aspects of an accommodative program that meet their particular needs. Furthermore, we find no evidence that high levels of program support are in themselves detrimental to functionally intact older people or that high levels of program autonomy are detrimental to functionally impaired residents. Rather, in an accommodative setting, individuals will selectively respond to those aspects of the setting that meet their needs.

IMPLICATIONS FOR PROGRAM EVALUATION AND DESIGN

Procedures for assessing the quality of residential facilities can be applied in program evaluation and design. As already noted, the MEAP can be used to assess the level of program implementation, describe program models and compare them with existing facilities, and examine the differences between preferences and the actual facility features. We illustrated some of these applications in chapters 3, 4, 5, and 11. We focus here on the use of the MEAP to monitor program change, to promote program improvement, and to help design new programs.

Monitoring Program Change

Residential programs change over time as a secondary result of external factors and as a result of intentional, directed efforts. The MEAP is useful for charting the life history of a program as it undergoes such change.

Change in Ownership

Because of an increasing emphasis on cost containment and economies of scale, many privately owned facilities are being sold to large proprietary corporations with centralized administrative personnel and procedures. We evaluated one such nursing home before and after it was sold to an out-of-state corporation. The corporate owner hired a new administrator and changed some policies and services. For example, even though the residents were relatively impaired, the nursing home decreased its stated acceptance of disability and of problem behavior and placed more emphasis on resident participation in decision making. Thus, the MEAP documented an increase in performance demands and a shift to formal input from residents.

The social climate also changed; both the residents and staff reported a drop in self-disclosure and resident influence. The shift from personalized management by an owner/administrator to the more formal management by a large corporation, the imposition of higher standards of independent functioning and conformity, and increased discrepancy between facility policies and residents' actual functioning all may have contributed to residents' increased reluctance to talk openly and to their reduced sense of control. The feeling of community that had developed over the years with a stable administration had given residents informal channels for expressing their feelings and influencing decisions. The new owner's addition of a house meeting and other formal avenues for influence did not compensate for the loss of communication with the administrative staff or for the imposition of more stringent behavioral standards.

In addition, staff reported that the facility was less cohesive, placed less emphasis on independence, and was more disorganized following the change in ownership. Staff complained about the change in administrative philosophy from resident- to profit-centered and about the accompanying overwork, shortages of supplies, and indifference to residents' and staff's needs.

Results of the evaluation came as a surprise to the administration but were in a form that could be assimilated and used as a basis for instituting some changes. The staff were relieved to be able to communicate their dissatisfaction as a group and in a more objective form. As in this example, the MEAP can alert administrators to unwanted changes in a facility before they become firmly established.

Relocation and Renovation

The MEAP can be used to describe changes that accompany relocation and associated interventions. For example, we used the PAF to indicate specific areas in which the physical environment changed when a nursing home was relocated to a newly constructed building (chapter 10). Such information can be shared with managers and staff and used to guide design and policy changes. For example, early feedback to staff about residents' lack of use of designated social spaces could have led to corrective action, such as assigning a staff member to the lounge or ensuring that less-mobile residents spent some time in the social areas. This might have promoted better adaptation to the larger and more complex layout of the new building.

Smith and Buckwalter (1990) studied the reorganization of a county home into therapeutic units, which required relocation of nearly all residents and staff, as well as a shift in treatment philosophy. To reduce the stress of relocation, unit meetings were held to discuss goals of the reorganization and plans for the move; nursing students offered supportive interventions, including social activities to help residents get to know each other; and residents discussed their preferences for roommates and helped move their belongings from one unit to the other. According to the SCES, resident and staff respondents reported an increase in cohesion and decrease in comfort from before to after the relocation. This information helped facility administrators understand the effects of the intervention and develop new program plans.

Program Enrichment

Berkowitz, Waxman, and Yaffe (1988) used the SCES to evaluate a congregate apartment with an enriched self-help program. The self-help program was based on the idea that when residents do more for themselves, their morale and well-being often improve. Residents were encouraged to participate in decision making, and new, professionally trained staff were added to the target housing unit; two other apartment buildings were comparison sites. As predicted, residents in the self-help program reported more cohesion, independence, and resident influence as well as higher self-esteem than did residents in the comparison sites.

Another way to improve program quality is to educate staff members in gerontological concepts and values. Following this idea, Spinner (1986) had staff members in two nursing homes participate in an 8-week educational program; staff in two other homes received no such training. Compared with staff in the control group, staff in the special program reported more emphasis on self-disclosure and resident influence following the program. However, residents did not perceive any changes. Staff development programs may improve staff members' views of facility social climate, but more intensive interventions and follow-up may be needed before such programs show an impact on residents (e.g., Stirling & Reid, 1992).

Promoting Program Improvement

Another use of the MEAP is to help formulate interventions in a facility. As a first step, implementation standards can be used to identify the need for change; for example, a discrepancy between a facility's environment and normative characteristics in other facilities may point to potentially desirable changes. Alternatively, program modifications may be guided by differences between facility characteristics and residents' or staff members' preferences, theoretical ideas and experts' opinions about how facilities should function, or empirical information about the influence of program characteristics on resident and staff outcomes.

Along these lines, Chambers and his colleagues (1988) developed guidelines to assess how well residential facilities conform to the principles underlying a community mental health promotion program. The researchers consulted the MEAP when formulating the guidelines, which cover the physical environment, prevention of behavior problems, communication, staff responses to residents' needs, and residents' activities in the facility and in the community. The guidelines are designed to be completed jointly by a facility manager and nurse-surveyor and to reflect an attainable therapeutic standard of care. They can be used to evaluate program implementation and to identify areas in need of improvement.

We recommend taking four steps when using the MEAP to plan change: assessment, feedback, planning and instituting change, and reassessment (Finney & Moos, 1984; Moos, 1988a). An organizational development project in two long-term care facilities illustrates this process (Wells & Singer, 1988; Wells, Singer, & Polgar, 1986). Initial assessment identified problems such as low expectations of resident functioning and policies that allowed residents little choice in their daily activities or participation in facility planning. After the findings were presented to staff, committees composed of residents, staff, and family members were organized to clarify important problem areas and to make changes. These committees planned for additional orientational aids, formed a subcommittee to welcome new residents and their families, prepared an informational brochure for new residents, and formed a self-help group to increase residents' confidence in public speaking.

Residents enjoyed the mental stimulation and companionship of the committees; they formed closer relationships and communicated more openly with each other. Residents also gained self-confidence and skills in presenting their ideas, and staff began to see them as more competent. Sharing the results of the initial MEAP assessment contributed to the organizational development project by serving as a stimulus for discussion and as a means of identifying areas in which to initiate change. In addition, the broad participation and open communication served as a model of the way the project should work and helped the project coordinators develop task-oriented contacts with residents, staff, and families.

More broadly, plans for program modifications can develop into social action. Thus, when Nelson and Earls (1986) identified serious problems in the quality of community housing for long-term psychiatric patients, they presented their findings in a community forum just prior to a local election. Service providers, patients, and family members elaborated on the survey data, and local political candidates pledged to form an action coalition to improve community housing options for long-term patients.

Guiding Program Design

Researchers and planners have developed a substantial amount of information on how to design residential facilities for impaired older people (Regnier & Pynoos, 1987). Here, we highlight some central issues that need to be addressed in formulating specific plans for facility design and programming (for a more detailed discussion, see Moos, Lemke, & David, 1987).

Enhancing Program Support

Supportive features in a program include physical design features aimed at making the environment safer and easier to negotiate, staffing characteristics, and services. These aspects of the facility are designed to meet basic life-maintenance needs and therefore have greatest salience for those whose functional limitations interfere with their ability to meet these needs independently. When designers and planners consider how far to extend support in a facility, they need to keep in mind existing economic constraints and the intrusiveness, stigma, and induced helplessness that may accompany support.

For individuals to exercise their capacities, a facility must assure residents of a minimum level of physical safety. Older people have strong preferences for features that increase their safety, such as smoke detectors in halls and rooms, call buttons in bathrooms and bedrooms, and nonslip surfaces on steps and bathroom floors. Because these basic safety features are a major focus of licensing and regulation efforts, nearly all facilities offer them. However, surveillance and risk reduction can diminish residents' privacy and impart an institutional atmosphere to the facility. For example, 24-hour physical surveillance provides the ultimate in security but is generally considered too intrusive except for extremely impaired, at-risk individuals.

The tension between support and autonomy is even more apparent in the case of prosthetic and orientational aids. Some older people view such features as stigmatizing. In addition, some older people resist supportive features and services because they wish to exclude frail and disabled older people from their residential setting. Thus, characteristics of users, rather than the features themselves, may be viewed as stigmatizing.

Nevertheless, most older people wish to age in place and can do so more easily when the environment provides support. Furthermore, a facility's inattention to prosthetic and orientational features may lead to reductions in residents' functioning and to negative self-labeling. For example, a floor that makes walking difficult or seats that make standing up a strain may lead older persons to view themselves as sick or frail. A complex floor plan with few aids to orientation can contribute to residents' confusion and discourage exploration. As we saw in our relocation study (chapter 10), removing physical barriers to access can result in important benefits for wheelchair-mobile residents. And some older people, particularly those with some impairment, indicate a strong desire for prosthetic features and orientational aids.

The appropriate staffing level in a facility is a function of factors such as facility size and the level of resident disability. In this case, the residents' needs for support and autonomy must be further balanced against staff stress and job performance. In facilities with higher staffing levels, the social climate emphasis on independence tends to be less. However, the background and training of staff, more than their actual numbers, may have a greater impact on facility functioning. Facilities with more diverse, better-trained staff tend to be more harmonious and place greater emphasis on resident autonomy.

Several guidelines can be drawn from this discussion: (1) Because supportive features may cause older persons to be stigmatized as different and less competent, they should be as unobtrusive as possible; (2) supportive features and services should be noncoercive, that is, they should be available to residents who need them but should not be imposed on the entire resident group; and (3) norms regarding supportive features are likely to change over time as public buildings, workplaces, and residences become more accessible to people with physical handicaps.

Promoting Choice and Control

Policies, design features, and staff behaviors all influence residents' autonomy, which is generally in dynamic tension with support. Designers and planners face several issues with regard to promoting choice and control. For both residents and staff, too much emphasis on autonomy can jeopardize order and predictability in the environment. Some areas of communal life are so emotionally charged that an emphasis on autonomy may lead to divisive conflict and threaten group cohesion.

To support autonomy in communal residential settings, the physical design must allow adequate space and amenities for various functions. We also believe that residents should be provided with maximum feasible privacy and control over their daily activities, even though this may make managing the facility somewhat harder. Most older people want choices concerning mealtimes, seating at meals, and their daily schedule. Many also want the opportunity to participate in house meetings and resident committees and to be involved in planning facility activities. As we have seen, residents, particularly those who are functionally able, tend to function better and be more active in facilities programmed to promote choice and control.

Encouraging Activity and Involvement

One important determinant of satisfaction in old age is continuity in one's pattern of engagement with others and of participation in meaningful activities. A facility can promote interaction by providing features to support social and recreational activities, carefully planning the location of communal areas, making formal programs available, and recognizing individual autonomy. In this area, the issues for planners relate to the appropriate boundaries between public and private space and to the active encouragement versus coercion of program participation.

To encourage social interaction and use, communal spaces should be located near heavily trafficked areas or staff workstations or have good views of outside activities. In an attempt to replicate the patterns found in private homes, designers sometimes distance social areas from caregiving areas. Although such designs may reduce the institutional quality of these settings, they raise some new problems. For example, the design of the new building in our relocation study encouraged staff to leave immobile residents in their bedrooms so as to avoid isolation from staff contact in the communal social areas or further crowding in the hallways around the nurses' stations. We found a similar reduction in use of social spaces for ambulatory residents. Thus, particularly for an impaired population, it may be important to recognize and support the need for staff–resident contact and to allow the boundaries between personal and social spaces to be somewhat blurred (Kayser-Jones, 1989), as in the innovative nursing home design evaluated by Lawton, Fulcomer, and Kleban (1984).

In addition to increasing physical accessibility and attractiveness of communal areas, planners need to consider staff behavior and program policies and services. For example, to encourage social interaction and use of communal areas, staff may need to transport and monitor immobile or confused residents. Encouraging residents to assist in transporting fellow residents may also be desirable, as are efforts to provide residents increased control over their own mobility, for example, using motorized wheelchairs.

A successful program to promote interaction among residents should (1) incorporate resident input in planning, (2) provide flexible rules for participation so that residents who are less inclined to socialize do not feel pressured, (3) emphasize community as well as facility activities to counteract insularity, (4) recognize the importance of privacy as an essential element in encouraging meaningful social interaction, and (5) concentrate special attention on physically and mentally impaired residents who need extra encouragement to remain active.

Accommodating Residents' Changing Needs

An environment in which physical features and services remain constant and do not allow for residents' increasing disabilities can cause considerable disruption when impaired residents must be transferred to another facility. Rather than requiring temporary relocation or premature institutionalization, administrators often accommodate residents' aging by allowing an extended period of residence despite increasing impairments (Lawton et al., 1985), administrators may simultaneously relax admission requirements. As a result, congregate facilities undergo an aging process as a function of changes among their residents. The aging of the resident population can eventually reduce the independent character of the facility and alter the population of new applicants. These changes in the resident population need to be accompanied by adjustments in physical features and services in order to maintain relative congruence between resident and facility characteristics.

One concern is that a protected, service-rich housing environment may enhance comfort and satisfaction among high-functioning older residents and thus undermine their independence. For example, Rutman and Freedman (1988) found that older people on a waiting list to enter congregate apartments feared loss of privacy and independence as a consequence of the move; however, among those who moved, adjustment and satisfaction increased. Similarly, Gutman (1978, 1988) found no evidence of differential decline in health status, activity level, or level of interaction with family and friends for older people moving into multilevel facilities as compared to those in housing with fewer services. We also found no negative outcomes associated with settings having more supportive features. Although there may be a potential for premature dependency, the availability of physical and program supports tends to enhance residents' feelings of security and enable them to enjoy independence by maintaining themselves at optimal levels.

FUTURE DIRECTIONS

The Assessment of Residential Programs

The Institute of Medicine (IOM, 1986) recommendations have led to important new developments in nursing home assessment that may generalize to other residential settings. The IOM report emphasized the need for ongoing assessment of residents' medical, functional, and psychosocial needs; residents' active participation in decisions about their care and facility policies; inclusion of information about the facility environment and residents' preferences and quality of life in certification reviews; and the development of process and outcome criteria of nursing home performance based on information obtained directly from residents and staff. Most of these recommendations were included in nursing home reform legislation enacted in the Omnibus Budget Reconciliation Act of 1987.

These developments have led to growing interest in enriching the conceptual and empirical basis of assessment in nursing homes and other types of residential care facilities. Thus, to specify the major components of nursing home quality, Glass (1991) presented a conceptual framework that considers the quality of staff-resident interactions, the personalization of residents' rooms, and the appeal of meals, as well as more traditional indices of the services provided and characteristics of the physical environment. Hannan, O'Donnell, and Lefkowich (1989) described a patient-centered quality assurance review system that includes information on residents' grooming and environment, meal services, and residents' participation in activities.

A new federal survey process, the Patient Care and Services (PaCS), reduces the focus on facility policies and procedures, increases the time spent on reviewing resident care, and mandates personal interviews with a sample of facility residents. In a randomized controlled evaluation of a pilot version of PaCS, Spector and Drugovich (1989) found that a PaCS survey team identified more problems than did a traditional survey team, but that there were no differences in resident outcomes in nursing homes surveyed by PaCS teams versus those surveyed by traditional teams. The authors noted the PaCS assessment may not be a powerful enough intervention to influence outcomes and that more specific direction is needed to link survey findings to potential facility modifications.

One way to emphasize the importance of the assessment process is to base reimbursement levels in part on facility characteristics and resident outcomes. In a first step toward implementing this idea, the Illinois Quality Incentive Program (QUIP) offers nursing homes a bonus payment for achievement of each of six quality standards (Geron, 1991). These standards focus on facility structure and environment, resident participation and choice, specialized intensive services, care plans, resident satisfaction, and community and family participation. Geron found that more than 90% of eligible facilities chose to participate in the program and that an increasing proportion of facilities was able to achieve bonus payments. The receipt of bonus payments was publicly recognized, so that a facility that met all six quality standards became known as a six-star facility. These findings show that systematic assessment and feedback of information about a facility, coupled with some financial and publicity incentives, can lead to improvements in the quality of care. (But see Chapin and Silloway [1992] and Geron [1991] for some cautions about the use of incentive systems based on outcomes.)

Procedures such as the MEAP can contribute to these efforts by expanding the content of certification assessments and focusing on commonalities in the underlying features of diverse types of residential programs. A consistent conceptual framework and assessment methodology for the entire range of facilities may be helpful to developing a better understanding of how they function. As another step in this direction, we have developed a modification of the MEAP, the Residential Psychiatric Programs Inventory (RESPPI), to assess the key characteristics of psychiatric and substance abuse treatment programs. Efforts such as these can help to develop general principles to understand how residential facilities affect residents and staff.

The Focus of Intervention and Evaluation

As befits a field in a state of growth, new program models and interventions are being implemented in congregate residential facilities. These include improving the staff training (Burgio & Burgio, 1990), instituting mental health programs in long-term care (Smyer, 1989), integrating supportive services in congregate housing facilities (Blandford, Chappell, & Marshall, 1989; Brice & Gorey, 1991), increasing opportunities for meaningful activity (Kautzer, 1988; Voeks & Drinka, 1990), and encouraging more family and community involvement in residential facilities. With so much flux in programs, evaluation research becomes all the more important. We focus here on three somewhat neglected issues that need to be considered in program evaluation.

First, personal variables other than functional ability may alter how facility characteristics affect adaptation and should be considered in formulating evaluations. These variables include personality characteristics and values as well as residents' approach and avoidance coping styles. As noted in our model (Figure 1.1), residents' coping responses in a particular setting are affected by objective environmental factors and facility social climate and by personal characteristics; in turn, coping efforts influence a variety of outcomes, including residents' adjustment, activity levels, and functional dependencies.

In a longitudinal study of older adults about to enter long-term care facilities, Kahana, Kahana, and Young (1987) asked individuals how they would cope with the impending relocation. Older people who relied more on problem-focused or approach coping and less on emotional discharge or avoidance coping prior to entry into long-term care showed higher morale and self-esteem and better cognitive functioning 9 months later. Similarly, Rodin (1983) found that nursing home residents who learned effective skills for coping with daily stressors showed increases in their feelings of control and in their activity level. Thus, in evaluating interventions that enrich the facility environment, it is important to recognize that residents' increased reliance on active problem solving may mediate these programs' positive influence on resident well-being. In addition, individuals who cope more actively may enter facilities that have policies oriented toward resident self-direction. These facilities may do more to support residents' approach coping strategies and, consequently, to improve their well-being.

Second, evaluators need to consider the effects of interventions on staff and other caretakers. Our findings suggest that conditions that contribute to a positive environment for residents, such as harmony, resident-direction, and good organization, also help improve staff morale and performance (chapter 8). In addition, allowing staff some respite from heavy patient care demands and developing a decentralized organization in which management seeks input from staff also seem to improve staff morale and facilitate individualized patterns of care (e.g., Stirling & Reid, 1992). Overall, the conceptual framework we have applied to residents can also be applied to help understand the determinants of staff work performance, which is a key contributor to residents' quality of life.

Finally, it is important to consider the context in which an intervention is implemented and to understand that residential programs are open systems shaped in part by external community forces. The effect of an intervention is likely to depend on the characteristics of the facility in which it takes place. For example, Head, Portnoy, and Woods (1990) found that reminiscence training had a positive effect on residents in a day-care center with relatively few competing activities but had no such effect in a center with a richer environment and many alternative activities. Similarly, Cherry (1991) found that an ombudsman program was associated with better quality of nursing care; however, this was only the case in nursing homes that had adequate numbers of registered nurses on the staff. Thus, to understand why a new program is or is not successful, evaluators need to assess comprehensively other aspects of the facility environment (Kahana, Kahana, & Riley, 1988).

The Theoretical Perspective

We have used a congruence or needs-hierarchy model to organize some complex relationships between personal and environmental factors and resident outcomes. In this section, we note some developments that are needed to enhance the model's use as a predictive tool. We also consider some new theoretical formulations and their implications for work in congregate residential programs.

To put the congruence model into operation, systematic measures of personal and environmental features are needed. Researchers have developed measures of personal attributes, and we have developed the MEAP specifically to assess congregate residential settings for the elderly. A congruence model also requires some degree of functional correspondence between individual competence and environmental conditions. For example, an individual's mobility status is logically connected with the physical accessibility of the environment. Similarly, an individual's mental status and the orientational aids in an environment are logically related; they are areas in which congruence effects might be expected. In other words, adding orientational aids to a building is likely to have a differential impact on residents who vary in mental status but not on those who vary in mobility status.

Having devised measures for both environmental and personal attributes, researchers need to examine how the scales in each domain correspond. For example, we have evidence that the new building we studied in our relocation project resulted in improvements in use, but we do not know whether it resulted in optimal use by either ambulatory or wheelchair-mobile residents. In addition, researchers should consider physical and social factors together. For example, increased accessibility can have a positive influence on immobile residents only if policies and staff behavior include efforts to move them out of their bedrooms into social spaces.

Another objective is to identify the range of environmental challenges that promotes competent residents' well-being without detracting from the well-being of impaired residents and the range of environmental supports that promotes impaired residents' functioning without undermining that of the more competent residents. These ranges will help to define the extent of resident heterogeneity that can be managed in one setting without detriment to one or another subgroup of residents.

Lawton (1990) has emphasized the need to focus more on environmental proactivity instead of environmental docility. Thus, the model also needs to be extended to reflect the fact that residents and staff actively try to transform their environment in order to maintain equilibrium with it. For example, program managers, responding to a decline in the average functional ability of residents in a congregate apartment, may try to develop an accommodative housing model with more prosthetic features and services. In the absence of social spaces adjacent to staff areas in a facility, staff and residents may transform the hallways into lounge areas, as they did in our relocation study. Conversely, when residents do not use specific services or activities, they are likely to be discontinued.

To inform the design of residential facilities as communities for growth, these ideas need to be integrated with recent theories of successful aging. According to continuity theory, older adults try to maintain some consistency in their psychological view of themselves and in their external circumstances (Atchley, 1989). When the environment changes, older adults try to find continuity by managing the new situation in familiar ways. However, congregate residential settings are by their very nature different from anything most older people have experienced. Superficial attempts to make congregate settings homelike may simply derail attempts to cope, but efforts to build new coping strategies, to provide privacy and meaningful personal choice, and to orient residents to the new environment may help maintain internal continuity.

Baltes and Baltes (1990) have presented a differentiated psychological model of successful aging that emphasizes selective optimization and compensation. Selection involves a reduction in the number of areas in which an older adult chooses to remain active. Optimization is the effort to maintain reasonably high levels of functioning in the chosen areas. Compensation involves the development of substitute skills and activities when an individual's competence declines and key environmental resources are no longer available. This may include a search for supportive aids and services and for more humanistic environments.

Residential facilities for older adults need to be planned and managed according to the principles of continuity, selective optimization, and compensation. With these principles as guidelines, and with improved assessment and evaluation programs to provide ongoing feedback to an informed and competent staff, older adults may be able to maintain their sense of meaning until the very end of life.

References

Achenbaum, W. A. (1978). *Old age in the new land.* Baltimore: Johns Hopkins University Press.

Adams, R. G. (1985–86). Emotional closeness and physical distance between friends: Implications for elderly women living in age-segregated and age-integrated settings. *International Journal of Aging and Human Development, 22,* 55–76.

Agbayewa, M. O., Ong, A., & Wilden, B. (1990). Empowering long term care facility residents using a resident staff group approach. *Clinical Gerontologist, 9,* 191–201.

Andersen, R., & Newman, J. F. (1973). Societal and individual determinants of medical care utilization in the United States. *Milbank Memorial Fund Quarterly, 51,* 95–124.

Antonucci, T. C. (1990). Social supports and social relationships. In R. H. Binstock & L. K. George (Eds.), *Handbook of aging and the social sciences* (3rd ed., pp. 205–226). San Diego: Academic Press.

Arling, G., Harkins, E. B., & Capitman, J. A. (1986). Institutionalization and personal control: A panel study of impaired older people. *Research on Aging, 8,* 38–56.

Arling, G., Nordquist, R. H., & Capitman, J. A. (1987). Nursing home cost and ownership type: Evidence of interaction effects. *Health Services Research, 22,* 255–269.

Arrington, B., & Haddock, C. C. (1990). Who really profits from not-for-profits? *Health Services Research, 25,* 291–304.

Atchley, R. C. (1989). A continuity theory of normal aging. *The Gerontologist, 29,* 183–190.

Avant, W. R., & Dressel, P. L. (1980). Perceiving needs by staff and elderly clients: The impact of training and client contact. *The Gerontologist, 20,* 71–77.

Avorn, J., & Langer, E. (1982). Induced disability in nursing home patients: A controlled trial. *Journal of the American Geriatrics Society, 30,* 397–400.

Babchuk, N., Peters, G. R., Hoyt, D. R., & Kaiser, M. A. (1979). The voluntary associations of the aged. *Journal of Gerontology, 34,* 579–587.

Baglioni, A. J. (1989). Residential relocation and health of the elderly. In K. S. Markides & C. L. Cooper (Eds.), *Aging, stress, and health* (pp. 119–137). New York: Wiley.

Bakos, M., Bozic, R., Chapin, D., & Neuman, S. (1980). Effects of environmental changes on elderly residents' behavior. *Hospital and Community Psychiatry, 31,* 677–682.

Baltes, M. M. (1988). The etiology and maintenance of dependency in the elderly: Three phases of operant research. *Behavior Therapy, 19,* 301–319.

Baltes, M. M., & Reisenzein, R. (1986). The social world in long-term care institutions: Psychosocial control toward dependency? In M. M. Baltes & P. B. Baltes (Eds.), *The psychology of control and aging* (pp. 315–343). Hillsdale, NJ: Erlbaum.

Baltes, M. M., & Wahl, H. (1992). The behavior system of dependency in the elderly: Interaction with the social environment. In M. G. Ory, R. P. Abeles, & P. D. Lipman (Eds.), *Aging, health, and behavior* (pp. 83–106). Newbury Park, CA: Sage.

Baltes, P. B., & Baltes, M. M. (1990). Psychological perspectives on successful aging: The model of selective optimization with compensation. In P. B. Baltes & M. M. Baltes (Eds.), *Successful aging: Perspectives from the behavioral sciences* (pp. 1–34). Cambridge: Cambridge University Press.

Banziger, G., & Roush, S. (1983). Nursing homes for the birds: A control-relevant intervention with bird feeders. *The Gerontologist, 23,* 527–531.

Barker, J. C., Mitteness, L. S., & Wood, S. J. (1988). Gate-keeping: Residential managers and elderly tenants. *The Gerontologist, 28,* 610–619.

Baum, A., & Paulus, P. B. (1987). Crowding. In D. Stokols & I. Altman (Eds.), *Handbook of environmental psychology* (Vol. 1, pp. 533–570). New York: Wiley.

Benjamin, L. C., & Spector, J. (1990). Environments for the dementing. *International Journal of Geriatric Psychiatry, 5,* 15–24.

Bennett, R. (Ed.). (1980). *Aging, isolation and resocialization.* New York: Van Nostrand Reinhold.

Bennett, R., & Nahemow, L. (1965). Institutional totality and criteria of social adjustment in residences for the aged. *Journal of Social Issues, 21,* 44–78.

Berdes, C. (1987). The modest proposal nursing home: Dehumanizing characteristics of nursing homes in memoirs of nursing home residents. *Journal of Applied Gerontology, 6,* 372–388.

Bergman, S., & Cibulski, I. O. (1981). Environment, culture and adaptation in congregate facilities: Perspectives from Israel. *The Gerontologist, 21,* 240–246.

Berkowitz, M. W., Waxman, R., & Yaffe, L. (1988). The effects of a resident self-help model on control, social involvement and self-esteem among the elderly. *The Gerontologist, 28,* 620–624.

Billingsley, J. D., & Batterson, C. T. (1986). Evaluating long-term care facilities: A field application of the MEAP. *The Journal of Long-Term Care Administration, 14,* 16–19.

Blake, R. (1985–86). Normalization and boarding homes: An examination of paradoxes. *Social Work in Health Care, 11,* 75–86.

Blake, R. (1987). The social environment of boarding homes. *Adult Foster Care Journal, 1,* 42–55.

Blandford, A., Chappell, N., & Marshall, S. (1989). Tenant resource coordinators: An experiment in supportive housing. *The Gerontologist, 29,* 826–829.

Booth, T. (1985). *Home truths: Old people's homes and the outcome of care.* Brookfield, VT: Gower.

Borup, J. H. (1983). Relocation mortality research: Assessment, reply, and the need to refocus on the issues. *The Gerontologist, 23,* 235–242.

Bourestom, N., & Tars, S. (1974). Alterations in life patterns following nursing home relocation. *The Gerontologist, 14,* 506–510.

Bowker, L. H. (1982). *Humanizing institutions for the aged.* Lexington, MA: D. C. Heath.

Branch, L. G., & Jette, A. M. (1982). Institutionalization among the aged. *American Journal of Public Health, 72,* 1373–1379.

Branch, L. G., & Ku, L. (1989). Transition probabilities to dependency, institutionalization, and death among the elderly over a decade. *Journal of Aging and Health, 1,* 370–408.

Braun, B. I. (1991). The effect of nursing home quality on patient outcome. *Journal of the American Geriatrics Society, 39,* 329–338.

Braun, K. L., & Rose, C. L. (1989). Goals and characteristics of long-term care programs: An analytic model. *The Gerontologist, 29,* 51–58.

Braun, K. L., Rose, C. L., & Finch, M. D. (1991). Patient characteristics and outcomes in institutional and community long-term care. *The Gerontologist, 31,* 648–656.

Brennan, P. L., & Moos, R. H. (1990). Physical design, social climate, and staff turnover in skilled nursing facilities. *The Journal of Long-Term Care Administration, 18,* 22–27.

Brennan, P. L., Moos, R. H., & Lemke, S. (1988). Preferences of older adults and experts for physical and architectural features of group living facilities. *The Gerontologist, 28,* 84–90.

Brennan, P. L., Moos, R. H., & Lemke, S. (1989). Preferences of older adults and experts for policies and services in group living facilities. *Psychology and Aging, 4,* 48–56.

Brent, R. S., Brent, E. E., & Mauksch, R. K. (1984). Common behavior patterns of residents in public areas of nursing homes. *The Gerontologist, 24,* 186–192.

Brice, G. C., & Gorey, K. M. (1991). Facilitating federally subsidized housing managerial role expansion: Beyond "bricks and mortar" to lifespace intervention with vulnerable older tenants. *Journal of Applied Gerontology, 10,* 486–498.

Brody, E. M. (1977). *Long-term care of older people: A practical guide.* New York: Human Sciences Press.

Brody, E. M. (1982). Service options in congregate housing. In R. D. Chellis, J. F. Seagle, & B. M. Seagle (Eds.), *Congregate housing for older people: A solution for the 1980s* (pp. 161–176). Lexington, MA: D. C. Heath.

Buffum, W. E., & Konick, A. (1982). Employees' job satisfaction, residents' functioning, and treatment progress in psychiatric institutions. *Health and Social Work, 7,* 320–327.

Burger, J. M. (1989). Negative reactions to increases in perceived personal control. *Journal of Personality and Social Psychology, 56,* 246–256.

Burgio, L. D., & Burgio, K. L. (1990). Institutional staff training and management: A review of the literature and a model for geriatric, long-term-care facilities. *International Journal of Aging and Human Development, 30,* 287–302.

Burgio, L. D., Engel, B. T., Hawkins, A., McCormick, K., & Scheve, A. (1990). A descriptive analysis of nursing staff behaviors in a teaching nursing home: Differences among NAs, LPNs, and RNs. *The Gerontologist, 30,* 107–112.

Butterfield, D., & Weidemann, S. (1987). Housing satisfaction of the elderly. In V. Regnier & J. Pynoos (Eds.), *Housing the aged: Design directives and policy considerations* (pp. 133–152). New York: Elsevier.

Caporael, L. R., Lukaszewski, M. P., & Culbertson, G. H. (1983). Secondary baby talk: Judgments by institutionalized elderly and their caregivers. *Journal of Personality and Social Psychology, 44,* 746–754.

Carp, F. M. (1978–79). Effects of the living environment on activity and use of time. *International Journal of Aging and Human Development, 9,* 75–91.

Carp, F. M. (1985). Relevance of personality traits to adjustment in group living situations. *Journal of Gerontology, 40,* 544–551.

Carp, F. M. (1987a). Environment and aging. In D. Stokols & I. Altman (Eds.), *Handbook of environmental psychology* (Vol. 1, pp. 329–360). New York: Wiley.

Carp, F. (1987b). The impact of planned housing: A longitudinal study. In V. Regnier & J. Pynoos (Eds.), *Housing the aged: Design directives and policy considerations* (pp. 43–79). New York: Elsevier.

Carp, F. M., & Carp. A. (1984). A complementary/congruence model of well-being or mental health for the community elderly. In I. Altman, M. P. Lawton, & J. F. Wohlwill (Eds.), *Elderly people and the environment* (pp. 279–336). New York: Plenum.

Carpman, J. R., Grant, M. A., & Simmons, D. A. (1986). *Design that cares: Planning health facilities for patients and visitors.* Chicago: American Hospital Publishing.

Carstensen, L. L. (1991). Selectivity theory: Social activity in life-span context. In K. W. Schaie (Ed.), *Annual review of geriatrics and gerontology* (Vol. 11, pp. 195–217). New York: Springer.

Caswell, R. J., & Cleverley, W. O. (1983). Cost analysis of the Ohio nursing home industry. *Health Services Research, 18,* 359–382.

Chambers, L., Forchuk, C., Munroe-Blum, H., Woodcox, V., Moore, G., & Wigmore, D. (1988). The development of guidelines to promote a therapeutic environment in lodging homes. *Canada's Mental Health,* December 1988, 14–18.

Chandler, J. T., Rachal, J. R., & Kazelskis, R. (1986). Attitudes of long-term care nursing personnel toward the elderly. *The Gerontologist, 26,* 551–555.

Chapin, R., & Silloway, G. (1992). Incentive payments to nursing homes based on quality-of-care outcomes. *Journal of Applied Gerontology, 11,* 131–145.

Cherry, R. L. (1991). Agents of nursing home quality of care: Ombudsmen and staff ratios revisited. *The Gerontologist, 3,* 302–308.

Christensen, D., & Cranz, G. (1987). Examining physical and managerial aspects of urban housing for the elderly. In V. Regnier & J. Pynoos (Eds.), *Housing the aged*: *Design directives and policy considerations* (pp. 105–132). New York: Elsevier.

Clark, H. M. (1989). A study of the relationships between personal characteristics, social environment and well-being in older nursing home residents (Doctoral dissertation, University of Maryland, College Park, 1988). *Dissertation Abstracts International, 50,* 1309B.

Clark, P. G. (1988). Autonomy, personal empowerment, and quality of life in long-term care. *Journal of Applied Gerontology, 7,* 279–297.

Coffman, T. L. (1981). Relocation and survival of institutionalized aged: A re-examination of the evidence. *The Gerontologist, 21,* 483–500.

Cohen, F., Bearison, D. J., & Muller, C. (1987). Interpersonal understanding in the elderly: The influence of age-integrated and age-segregated housing. *Research on Aging, 9,* 79–100.

Cohen, M. A., Tell, E. J., & Wallack, S. S. (1986). Client-related risk factors of nursing home entry among elderly adults. *Journal of Gerontology, 41,* 785–792.

Cohen, U., & Weisman, G. D. (1991). *Holding on to home: Designing environments for people with dementia.* Baltimore: Johns Hopkins University Press.

Cohen-Mansfield, J., Marx, M. S., & Werner, P. (1992). Observational data on time use and behavior problems in the nursing home. *Journal of Applied Gerontology, 11,* 111–121.

Cole, S. M. (1982). The social environments of nursing homes and their consequences for the styles of participation of older residents (Doctoral dissertation, Louisiana State University, Baton Rouge, 1981). *Dissertation Abstracts International, 42,* 4936A.

Colling, J. (1985). Nursing home residents' control of activities of daily living and well-being (Doctoral dissertation, Oregon Health Sciences University, Portland). *Dissertation Abstracts International, 46,* 3389B.

Conrad, K. J., Hanrahan, P., & Hughes, S. L. (1988). The use of profiles and models in evaluating program environments. In K. J. Conrad & C. Roberts-Gray (Eds.), *Evaluating program environments*: *New directions for program evaluation* (No. 40, pp. 25–43). San Francisco: Jossey-Bass.

Constable, J. F., & Russell, D. W. (1986). The effect of social support and the work environment upon burnout among nurses. *Journal of Human Stress, 12,* 20–26.

Conway, T. L., Vickers, R. R., & French, J. R. P. (1992). An application of person-environment fit theory: Perceived versus desired control. *Journal of Social Issues, 48(2),* 95–107.

Coppola, D., Feldheim, I., Kennaley, J., & Steinberg, A. (1990). Developing a sense of community: A programming approach for institutionalized elderly. *Activities, Adaptation and Aging, 14,* 17–25.

Coulton, C. J., Holland, T. P., & Fitch, V. (1984). Person-environment congruence and psychiatric patient outcome in community care homes. *Administration in Mental Health, 12,* 71–88.

Cranz, G. (1987). Evaluating the physical environment: Conclusions from eight housing projects. In V. Regnier & J. Pynoos (Eds.), *Housing the aged*: *Design directives and policy considerations* (pp. 81–104). New York: Elsevier.

Culhane, D. P., & Hadley, T. R. (1992). The discriminating characteristics of for-profit versus not-for-profit freestanding psychiatric inpatient facilities. *Health Services Research, 27,* 178–194.

Curry, T. J., & Ratliff, B. W. (1973). The effects of nursing home size on resident isolation and life satisfaction. *The Gerontologist, 13,* 295–298.

Cutler, S. J., & Hendricks, J. (1990). Leisure and time use across the life course. In R. H. Binstock & L. K. George (Eds.), *Handbook of aging and the social sciences* (3rd ed., pp. 169–185). San Diego: Academic Press.

Davis, M. A. (1991). On nursing home quality: A review and analysis. *Medical Care Review, 48,* 129–166.

Dennis, L. C., Burke, R. E., & Garber, K. G. (1977). Quality Evaluation System: An approach for patient assessment. *Journal of Long-Term Care Administration, 5,* 28–51.

Department of Veterans Affairs. (1991). *1990 annual report.* Washington, DC: U. S. Government Printing Office.

Deutschman, M. (1982). Environmental settings and environmental competence. *Gerontology and Geriatrics Education, 2,* 237–242.

Dolinsky, A. L., & Rosenwaike, I. (1988). The role of demographic factors in the institutionalization of the elderly. *Research on Aging, 10,* 235–257.

Duffy, M., Bailey, S., Beck, B., & Barker, D. G. (1986). Preferences in nursing home design: A comparison of residents, administrators, and designers. *Environment and Behavior, 18,* 246–257.

Dunlop, B. D. (1979). *The growth of nursing home care.* Lexington, MA: D. C. Heath.

Earls, M., & Nelson, G. (1988). The relationship between long-term psychiatric clients' psychological well-being and their perceptions of housing and social support. *American Journal of Community Psychology, 16,* 279–293.

Elston, J. M., Koch, G. G., & Weissert, W. G. (1991). Regression-adjusted small area estimates of functional dependency in the noninstitutionalized American population age 65 and over. *American Journal of Public Health, 81,* 335–343.

Elwell, F. (1984). The effects of ownership on institutional services. *The Gerontologist, 24,* 77–83.

Fernandez-Ballesteros, R., Diaz, P., Izal, M., & Gonzalez, J. L. (1987). Evaluacion de una residencia de ancianos y valoracion de intervenciones ambientales. In R. Fernandez-Ballesteros (Ed.), *El ambiente analisis psicologico* (pp. 227–248). Madrid: Don Ramon de la Cruz.

Fernandez-Ballesteros, R., Izal, M., Diaz, P., Gonzalez, J. L., Vila, E., & Espinosa, M. J. (1986). Estudio ecopsicologico de una residencia de ancianos. In R. Fernandez-Ballesteros (Ed.), *Evaluacion de contextos* (pp. 59–103). Murcia, Spain: Servicio de Publicaciones, Universidad de Murcia.

Fernandez-Ballesteros, R., Izal, M., Montorio, I., Llorente, M. G., Hernandez, J. M., & Guerrero, M. A. (1991). Evaluation of residential programs for the elderly in Spain and the United States. *Evaluation Practice, 12,* 159–164.

Finney, J. W., Mitchell, R. E., Cronkite, R. C., & Moos, R. H. (1984). Methodological issues in estimating main and interactive effects: Examples from coping/social support and stress field. *Journal of Health and Social Behavior, 25,* 85–98.

Finney, J. W., & Moos, R. H. (1984). Environmental assessment and evaluation research: Examples from mental health and substance abuse programs. *Evaluation and Program Planning, 7,* 151–167.

Finney, J. W., & Moos, R. H. (1992). The long-term course of treated alcoholism: II. Predictors and correlates of 10-year functioning and mortality. *Journal of Studies on Alcohol, 53,* 142–153.

Fottler, M. D., Smith, H. L., & James, W. L. (1981). Profits and patient care quality in nursing homes: Are they compatible? *The Gerontologist, 21,* 532–538.

Garritson, S. H. (1986). The influence of psychiatric inpatient environments on ethical decision making of psychiatric nurses (Doctoral dissertation, University of California, San Francisco, 1985). *Dissertation Abstracts International, 46,* 2622B.

Gelwicks, L. E., & Dwight, M. B. (1982). Programming for alternatives and future models. In R. D. Chellis, J. F. Seagle, & B. M. Seagle (Eds.), *Congregate housing for older people: A solution for the 1980s* (pp. 69–87). Lexington, MA: Lexington Books.

George, L. K. (1978). The impact of personality and social status factors upon levels of activity and psychological well-being. *Journal of Gerontology, 33,* 840–847.

Geron, S. M. (1991). Regulating the behavior of nursing homes through positive incentives: An analysis of Illinois' Quality Incentive Program (QUIP). *The Gerontologist, 31,* 292–301.

Glass, A. P. (1991). Nursing home quality: A framework for analysis. *Journal of Applied Gerontology, 10,* 5–18.

Goffman, E. (1961). *Asylums: Essays on the social situation of mental patients and other inmates.* Garden City, NY: Doubleday.

Golant, S. M. (1986). Subjective housing assessments by the elderly: A critical information source for planning and program evaluation. *The Gerontologist, 26,* 122–127.

Gottesman, L. E., Peskin, E., & Kennedy, K. M. (1990). Research and program experience in residential care facilities: Implications for mental health services to elderly and middle-aged clients. In E. Light & B. Lebowitz (Eds.), *The elderly with chronic mental illness* (245–281). New York: Springer.

Grant, P. R., Skinkle, R. R., & Lipps, G. (1992). The impact of an interinstitutional relocation on nursing home residents requiring a high level of care. *The Gerontologist, 32,* 834–842.

Grau, L., Chandler, B., Burton, B., & Kolditz, D. (1991). Institutional loyalty and job satisfaction among nurse aides in nursing homes. *Journal of Aging and Health, 3,* 47–65.

Green, I., Fedewa, B. E., Johnston, C. A., Jackson, W. M., & Deardorff, H. L. (1975). *Housing for the elderly: The development and design process.* New York: Van Nostrand Reinhold.

Greene, V. L., & Monahan, D. J. (1981). Structural and operational factors affecting quality of patient care in nursing homes. *Public Policy, 29,* 399–415.

Greene, V. L., & Ondrich, J. I. (1990). Risk factors for nursing home admissions and exits: A discrete-time hazard function approach. *Journal of Gerontology:Social Sciences, 45,* S250-S258.

Greenwald, S. R., & Linn, M. W. (1971). Intercorrelations of data on nursing homes. *The Gerontologist, 11,* 337–340.

Gustafson, D. H., Sainfort, F. C., Van Konigsveld, R., & Zimmerman, D. R. (1990). The Quality Assessment Index (QAI) for measuring nursing home quality. *Health Services Research, 25,* 97–127.

Gutman, G. M. (1978). Issues and findings relating to multilevel accommodation for seniors. *Journal of Gerontology, 33,* 592–600.

Gutman, G. M. (1988). The multilevel, multiservice model. In G. M. Gutman & N. K. Blackie (Eds.), *Housing the very old* (pp. 136–160). Burnaby, British Columbia, Canada: Simon Fraser University, Gerontology Research Centre.

Hannan, E. L., O'Donnell, J. F., & Lefkowich, W. K. (1989). An evaluation of the quality assurance system for skilled nursing facilities in New York State. *Evaluation and the Health Professions, 12,* 235–254.

Harel, Z. (1981). Quality of care, congruence and well-being among institutionalized aged. *The Gerontologist, 21,* 523–531.

Hartman, C., Horovitz, J., & Herman, R. (1987). Involving older persons in designing housing for the elderly. In V. Regnier & J. Pynoos (Eds.), *Housing the aged: Design directives and policy considerations* (pp. 153–175). New York: Elsevier.

Hatcher, M., Gentry, R., Kunkel, M., & Smith, G. (1983). Environmental assessment and community intervention: An application of the social ecology model. *Psychosocial Rehabilitation Journal, 7,* 22–28.

Hawes, C., & Phillips, C. D. (1986). The changing structure of the nursing home industry and the impact of ownership on quality, cost, and access. In B. H. Gray (Ed.), *For-profit enterprise in health care* (pp. 492–541). Washington, DC: National Academy Press.

Head, D. M., Portnoy, S., & Woods, R. T. (1990). The impact of reminiscence groups in two different settings. *International Journal of Geriatric Psychiatry, 5,* 295–302.

Hewitt, S. M., LeSage, J., Roberts, K. L., & Ellor, J. R. (1985). Process auditing in long term care facilities. *Quality Review Bulletin, 11,* 6–15.

Hiatt, L. G. (1982). Grouping elders of different abilities. In R. D. Chellis, J. F. Seagle, & B. M. Seagle (Eds.), *Congregate housing for older people: A solution for the 1980s* (pp. 27–49). Lexington, MA: D. C. Heath.

Hiatt, L.G. (1987). Designing for the vision and hearing impairments of the elderly. In V. Regnier & J. Pynoos (Eds.), *Housing the aged: Design directives and policy considerations* (pp. 341–371). New York: Elsevier.

Hing, E. (1987). Use of nursing homes by the elderly: Preliminary data from the 1985 National Nursing Home Survey. *Advance data from vital and health statistics,* (No. 135)(DHHS Pub. No. PHS 87–1250.) Hyattsville, MD: Public Health Service, National Center for Health Statistics.

Hinrichsen, G. A. (1985). The impact of age-concentrated, publicly assisted housing on older people's social and emotional well-being. *Journal of Gerontology, 40,* 758–760.

Hochschild, A. R. (1973). *The unexpected community.* Englewood Cliffs, NJ: Prentice-Hall.

Hodge, G. (1984). *Shelter and services for the small town elderly: The role of assisted housing.* Kingston, Ontario, Canada: Queen's University at Kingston, School of Urban and Regional Planning.

Hodge, G. (1987). Assisted housing for Ontario's rural elderly: Shortfalls in product and location. *Canadian Journal on Aging, 6,* 141–154.

Holder, E. L., Frank, B. W., & Spalding, J. (1985). *A consumer perspective on quality care: The residents' point of view.* Washington, DC: The National Citizens Coalition for Nursing Home Reform.

Holland, T. P., Konick, A., Buffum, W., Smith, M. K., & Petchers, M. (1981). Institutional structure and resident outcomes. *Journal of Health and Social Behavior, 22,* 433–444.

Holmberg, R. H., & Anderson, N. N. (1968). Implications of ownership for nursing home care. *Medical Care, 7,* 300–307.

Hulicka, I. M., Morganti, J. B., & Cataldo, J. F. (1975). Perceived latitude of choice of institutionalized and noninstitutionalized elderly women. *Experimental Aging Research, 1,* 27–39.

Hunt, M. E., & Pastalan, L. A. (1987). Easing relocation: An environmental learning process. In V. Regnier & J. Pynoos (Eds.), *Housing the aged: Design directives and policy considerations* (pp. 421–440). New York: Elsevier.

Institute of Medicine Committee on Nursing Home Regulation. (1986). *Improving the quality of care in nursing homes.* Washington, DC: National Academy Press.

Izal, M. (1992). Residential facilities for older adults: Cross-cultural environmental assessment. *European Journal of Psychological Assessment, 8,* 118–134.

Johnson, C. L., & Grant, L. A. (1985). *The nursing home in American society.* Baltimore: Johns Hopkins University Press.

Johnson, P. D. (1981). Effects of increased personal and interpersonal control upon the well-being of institutionalized geriatrics (Doctoral dissertation, Arizona State University, Phoenix). *Dissertation Abstracts International, 42,* 748B.

Jones, J. P., & Batterson, C. T. (1982). *Quality of life survey.* Carmichael, CA: Eskaton Administrative Center.

Kahana, B., & Kahana, E. (1970). Changes in mental status of elderly patients in age-integrated and age-segregated hospital milieus. *Journal of Abnormal Psychology, 75,* 177–181.

Kahana, E. (1982). A congruence model of person-environment interaction. In M. P. Lawton, P. G. Windley, & T. O. Byerts (Eds.), *Aging and the environment: Theoretical approaches* (pp. 97–121). New York: Springer.

Kahana, E., Kahana, B., & Riley, K. P. (1988). Contextual issues in quantitative studies of institutional settings for the aged. In S. Reinharz & G. D. Rowles (Eds.), *Qualitative Gerontology* (pp. 197–216). New York: Springer.

Kahana, E. E., Kahana, B., & Young, R. (1987). Strategies of coping and postinstitutional outcomes. *Research on Aging, 9,* 182–199.

Kahana, E., Liang, J., & Felton, B. J. (1980). Alternative models of person-environment fit: Prediction of morale in three homes for the aged. *Journal of Gerontology, 35,* 584–595.

Kane, R. A., & Kane, R. L. (1987). *Long-term care: Principles, programs, and policies.* New York: Springer.

Kart, C. S., & Palmer, N. M. (1987). How do we explain differences in the level of care received by the institutionalized elderly? *Journal of Applied Gerontology, 6,* 53–66.

Kasl, S. V., & Rosenfield, S. (1980). The residential environment and its impact on the mental health of the aged. In J. E. Birren & R. B. Sloane (Eds.),

Handbook of mental health and aging (pp. 468–498). Englewood Cliffs, NJ: Prentice-Hall.

Kastenbaum, R. (1985). Dying and death: A life-span approach. In J. E. Birren & K. W. Schaie (Eds.), *Handbook of the psychology of aging* (2nd ed., pp. 619–643). New York: Van Nostrand Reinhold.

Kautzer, K. (1988). Empowering nursing home residents: A case study of "Living Is For the Elderly," an activist nursing home organization. In S. Reinharz & G. D. Rowles (Eds.), *Qualitative Gerontology* (pp. 163–183). New York: Springer.

Kayser-Jones, J. (1989). The environment and quality of care in long-term care institutions. In *Indices of quality in long-term care: Research and practice*. New York: National League for Nursing.

Kemper, P., & Murtaugh, C. M. (1991). Lifetime use of nursing home care. *New England Journal of Medicine, 324,* 595–600.

Kleemeier, R. W. (Ed.). (1961). *Aging and leisure*. New York: Oxford University Press.

Kodama, K. (1986). The development and the effectiveness of a rating scale for environmental features at residential facilities for the aged. *Journal of Architecture, Planning, and Environmental Engineering, 366,* 53–60.

Kodama, K. (1988a). The analysis of architectural complaints related to architectural conditions of the homes for the aged rated by the residents. *Journal of Architecture, Planning, and Environmental Engineering, 385,* 53–63.

Kodama, K. (1988b). How architectural conditions affect the morale and the environmental distress of the residents of old age homes. *Journal of Architecture, Planning, and Environmental Engineering, 390,* 77–85.

Koncelik, J. A. (1976). *Designing the open nursing home*. Stroudsburg, PA: Dowden, Hutchinson, & Ross.

Koncelik, J. A. (1982). Elements of geriatric design: The personal environment. In R. D. Chellis, J. F. Seagle, & B. M. Seagle (Eds.), *Congregate housing for older people: A solution for the 1980s* (pp. 139–158). Lexington, MA: Lexington Books.

Koncelik, J. A. (1987). Product and furniture design for the chronically impaired elderly. In V. Regnier & J. Pynoos (Eds.), *Housing the aged: Design directives and policy considerations* (pp. 373–398). New York: Elsevier.

Kosberg, J. I., & Tobin, S. S. (1972). Variability among nursing homes. *The Gerontologist, 12,* 214–219.

Kruzich, J. M., Clinton, J. F., & Kelber, S. T. (1992). Personal and environmental influences on nursing home satisfaction. *The Gerontologist, 32,* 342–350.

Langer, E. J., & Rodin, J. (1976). The effects of choice and enhanced personal responsibility for the aged: A field experiment in an institutional setting. *Journal of Personality and Social Psychology, 34,* 191–198.

Larson, R. (1978). Thirty years of research on the subjective well-being of older Americans. *Journal of Gerontology, 33,* 109–125.

Lawton, M. P. (1975). *Planning and managing housing for the elderly*. New York: Wiley.

Lawton, M. P. (1981). Sensory deprivation and the effect of the environment on management of the patient with senile dementia. In N. E. Miller & G. D. Cohen (Eds.), *Clinical aspects of Alzheimer's disease and senile dementia* (pp. 227–251). New York: Raven.

Lawton, M. P. (1982). Competence, environmental press, and the adaptation of older people. In M. P. Lawton, P. G. Windley, & T. O. Byerts (Eds.), *Aging and the environment: Theoretical approaches* (pp. 33–59). New York: Springer.

Lawton, M. P. (1985a). Activities and leisure. In M. P. Lawton & G. L. Maddox (Eds.), *Annual review of gerontology and geriatrics* (Vol. 5, pp. 127–164). New York: Springer.

Lawton, M. P. (1985b). Housing and living environments of older people. In R. H. Binstock & E. Shanas (Eds.), *Handbook of aging and the social sciences* (2nd ed., pp. 450–478). New York: Van Nostrand Reinhold.

Lawton, M. P. (1989). Behavior-relevant ecological factors. In K. W. Schaie & C. Schoder (Eds.), *Social structure and aging: Psychological processes* (pp. 57–78). Hillsdale, NJ: Erlbaum.

Lawton, M. P. (1990). Residential environment and self-directedness among older people. *American Psychologist, 45,* 638–640.

Lawton, M. P., Altman, I., & Wohlwill, J. F. (1984). Dimensions of environment-behavior research: Orientations to place, design, process, and policy. In I. Altman, M. P. Lawton, & J. F. Wohlwill (Eds.), *Elderly people and the environment* (pp. 1–15). New York: Plenum Press.

Lawton, M. P., Fulcomer, M., & Kleban, M. (1984). Architecture for the mentally impaired elderly. *Environment and Behavior, 16,* 730–757.

Lawton, M. P., Greenbaum, M., & Liebowitz, B. (1980). The lifespan of housing environments for the aging. *The Gerontologist, 20,* 56–64.

Lawton, M. P., Moss, M., & Fulcomer, M. (1986–87). Objective and subjective uses of time by older people. *International Journal of Aging and Human Development, 24,* 171–188.

Lawton, M. P., Moss, M., & Grimes, M. (1985). The changing service needs of older tenants in planned housing. *The Gerontologist, 25,* 258–264.

Lawton, M. P., Nahemow, L., & Teaff, J. (1975). Housing characteristics and the well-being of elderly tenants in federally assisted housing. *Journal of Gerontology, 30,* 601–607.

Lawton, M. P., Patnaik, B., & Kleban, M. H. (1976). The ecology of adaptation to a new environment. *International Journal of Aging and Human Development, 7,* 15–26.

Lee, A. J., & Birnbaum, H. (1983). The determinants of nursing home operating costs in New York State. *Health Services Research, 18,* 285–308.

Lehman, A. F. (1983). The well-being of chronic mental patients: Assessing their quality of life. *Archives of General Psychiatry, 40,* 369–373.

Lehman, A. F. (1988). A quality of life interview for the chronically mentally ill. *Evaluation and Program Planning, 11,* 51–62.

Lehman, A. F., Possidente, S., & Hawker, F. (1986). The quality of life of chronic patients in a state hospital and in community residences. *Hospital and Community Psychiatry, 37,* 901–907.

Lemke, S., & Moos, R. H. (1980). Assessing the institutional policies of sheltered care settings. *Journal of Gerontology, 35,* 96–107.

Lemke, S., & Moos, R. H. (1981). The suprapersonal environments of sheltered care settings. *Journal of Gerontology, 36,* 233–243.

Lemke, S., & Moos, R. H. (1984). Coping with an intra-institutional relocation: Behavioral change as a function of residents' personal resources. *Journal of Environmental Psychology, 4,* 137–151.

Lemke, S., & Moos, R. H. (1985). The evaluation process in housing for the elderly. In G. M. Gutman & N. K. Blackie (Eds.), *Innovations in housing and living arrangements for seniors* (pp. 227–254). Burnaby, British Columbia, Canada: Simon Fraser University, Gerontology Research Centre.

Lemke, S., & Moos, R. H. (1986). Quality of residential settings for elderly adults. *Journal of Gerontology, 41,* 268–276.

Lemke, S., & Moos, R. H. (1987). Measuring the social climate of congregate residences for older people: Sheltered Care Environment Scale. *Psychology and Aging, 2,* 20–29.

Lemke, S., & Moos, R. H. (1989a). Ownership and quality of care in residential facilities for the elderly. *The Gerontologist, 29,* 209–215.

Lemke, S., & Moos, R. H. (1989b). Personal and environmental determinants of activity involvement among elderly residents of congregate facilities. *Journal of Gerontology: Social Sciences, 44,* S139-S148.

Lemke, S., & Moos, R. H. (1990). Validity of the Sheltered Care Environment Scale: Conceptual and methodological issues. *Psychology and Aging, 5,* 569–571.

Levey, S., Ruchlin, H. S., Stotsky, B. A., Kinloch, D. R., & Oppenheim, W. (1973). An appraisal of nursing home care. *Journal of Gerontology, 28,* 222–228.

Lieberman, M. A., & Tobin, S. S. (1983). *The experience of old age: Stress, coping, and survival.* New York: Basic Books.

Linn, M. W. (1966). A nursing home rating scale. *Geriatrics, 21,* 188–192.

Linn, M., Gurel, L., & Linn, B. (1977). Patient outcome as a measure of quality of nursing home care. *American Journal of Public Health, 67,* 337–344.

Lyman, K. A. (1989). Day care for persons with dementia: The impact of the physical environment on staff stress and quality of care. *The Gerontologist, 29,* 557–560.

Lyman, K.A. (1990). Staff stress and treatment of clients in Alzheimer's care: A comparison of medical and non-medical day care programs. *Journal of Aging Studies, 4,* 61–79.

Majone, G. (1984). Professionalism and nonprofit organizations. *Journal of Health Politics, Policy and Law, 8,* 639–659.

Maloney, N., & Bowman, R. (1982). *Riverview Hospital Social Learning Program: Description and evaluation* (Project Report). Port Coquitlam, Vancouver, British Columbia: Riverview Hospital, Department of Psychology.

Manning, N. (1989). *The therapeutic community movement: Charisma and routinisation.* London: Routledge.

Markson, E. W. (1982). Placement and location: The elderly and congregate care. In R. D. Chellis, J. F. Seagle, & B. M. Seagle (Eds.), *Congregate housing for older people: A solution for the 1980s* (pp. 51–65). Lexington, MA: D. C. Heath.

Mather, J. H. (1984). An overview of the Veterans Administration and its services for older veterans. In T. Wetle & J. Rowe (Eds.), *Older veterans: Linking VA and community resources* (pp. 35–48). Cambridge: Harvard University Press.

McAuley, W. J., & Arling, G. (1984). Use of in-home care by very old people. *Journal of Health and Social Behavior, 25,* 54–64.

McKay, N. L. (1991). The effect of chain ownership on nursing home costs. *Health Services Research, 26,* 109–124.

Meiners, M. R. (1982). An econometric analysis of the major determinants of nursing home costs in the United States. *Social Science and Medicine, 16,* 887–898.

Merrill, J., & Hunt, M. E. (1990). Aging in place: A dilemma for retirement housing administrators. *Journal of Applied Gerontology, 9,* 60–76.

Meyer, M. H. (1991). Assuring quality of care: Nursing home resident councils. *Journal of Applied Gerontology, 10,* 103–116.

Mirotznik, J., & Ruskin, A. P. (1985). Inter-institutional relocation and its effects on psychosocial status. *The Gerontologist, 25,* 265–270.

Moore, G. T., Tuttle, D. P., & Howell, S. C. (1985). *Environmental design research directions: Process and prospects.* New York: Praeger.

Moos, R. H. (1974). *Evaluating treatment environments: A social ecological approach.* New York: Wiley.

Moos, R. H. (1981). Environmental choice and control in community care settings for older people. *Journal of Applied Social Psychology, 11,* 23–43.

Moos, R. H. (1987). *The Social Climate Scales: A user's guide.* Palo Alto, CA: Consulting Psychologists Press.

Moos, R. H. (1988a). Assessing the program environment: Implications for program evaluation and design. In K. J. Conrad & C. Roberts-Gray (Eds.), *New directions for program evaluation: Evaluating program environments* (Vol. 40, pp. 7–23). San Francisco: Jossey-Bass.

Moos, R. H. (1988b). *Community-Oriented Programs Environment Scale manual* (2nd ed.). Palo Alto, CA: Consulting Psychologists Press.

Moos, R. H. (1989). *Ward Atmosphere Scale manual* (2nd ed.). Palo Alto, CA: Consulting Psychologists Press.

Moos, R. H., & Bromet, E. (1978). Relation of patient attributes to perceptions of the treatment environment. *Journal of Consulting and Clinical Psychology, 46,* 350–351.

Moos, R. H., David, T. G., Lemke, S., & Postle, E. (1984). Coping with an intrainstitutional relocation: Changes in resident and staff behavior patterns. *The Gerontologist, 24,* 495–502.

Moos, R. H., Gauvain, M., Lemke, S., Max, W., & Mehren, B. (1979). Assessing the social environments of sheltered care settings. *The Gerontologist, 19,* 74–82.

Moos, R., & Igra, A. (1980). Determinants of the social environments of sheltered care settings. *Journal of Health and Social Behavior, 21,* 88–98.

Moos, R. H., & Lemke, S. (1980). Assessing the physical and architectural features of sheltered care settings. *Journal of Gerontology, 35,* 571–583.

Moos, R. H., & Lemke, S. (1983). Assessing and improving social-ecological settings. In E. Seidman (Ed.), *Handbook of social intervention* (pp. 143–162). Beverly Hills, CA: Sage.

Moos, R. H., & Lemke, S. (1984). Supportive residential settings for older people. In I. Altman, M. P. Lawton, & J. F. Wohlwill (Eds.), *Elderly people and the environment* (pp. 159–190). New York: Plenum.

Moos, R. H., & Lemke, S. (1992a). *Multiphasic Environmental Assessment Procedure: A user's guide.* Palo Alto, CA: Department of Veterans Affairs and Stanford University Medical Center, Center for Health Care Evaluation.

Moos, R. H., & Lemke, S. (1992b). *Physical and Architectural Features Checklist manual.* Palo Alto, CA: Department of Veterans Affairs and Stanford University Medical Center, Center for Health Care Evaluation.

Moos, R. H., & Lemke, S. (1992c). *Policy and Program Information Form manual.* Palo Alto, CA: Department of Veterans Affairs and Stanford University Medical Center, Center for Health Care Evaluation.

Moos, R. H., & Lemke, S. (1992d). *Rating Scale manual.* Palo Alto, CA: Department of Veterans Affairs and Stanford University Medical Center, Center for Health Care Evaluation.

Moos, R. H., & Lemke, S. (1992e). *Resident and Staff Information Form manual.* Palo Alto, CA: Department of Veterans Affairs and Stanford University Medical Center, Center for Health Care Evaluation.

Moos, R. H., & Lemke, S. (1992f). *Sheltered Care Environment Scale manual.* Palo Alto, CA: Department of Veterans Affairs and Stanford University Medical Center, Center for Health Care Evaluation.

Moos, R. H., Lemke, S., & Clayton, J. (1983). Comprehensive assessment of residential programs: A means of facilitating evaluation and change. *Interdisciplinary Topics in Gerontology, 17,* 69–83.

Moos, R. H., Lemke, S., & David, T. G. (1987). Priorities for design and management in residential settings for the elderly. In V. Regnier & J. Pynoos (Eds.), *Housing the aged: Design directives and policy considerations* (pp. 179–205). New York: Elsevier.

Moos, R. H., & Schaefer, J. A. (1987). Evaluating health care work settings: A holistic conceptual framework. *Psychology and Health, 1,* 97–122.

Mor, V., Sherwood, S., & Gutkin, C. E. (1984). Psychiatric history as a barrier to residential care. *Hospital and Community Psychiatry, 35,* 368–372.

Mor, V., Sherwood, S., & Gutkin, C. (1986). A national study of residential care for the aged. *The Gerontologist, 26,* 405–417.

Morgan, A., & Godbey, G. (1978). The effect of entering an age-segregated environment upon the leisure activity patterns of older adults. *Journal of Leisure Research, 10,* 177–190.

Morgan, D. L. (1985). Nurses' perceptions of mental confusion in the elderly: Influences of resident and setting characteristics. *Journal of Health and Social Behavior, 26,* 102–112.

Mutran, E., & Ferraro, K. F. (1988). Medical need and use of services among older men and women. *Journal of Gerontology: Social Sciences, 43,* S162-S171.

Namazi, K. H., Eckert, J. K., Kahana, E., & Lyon, S. M. (1989). Psychological well-being of elderly board and care residents. *The Gerontologist, 29,* 511–516.

Nasar, J. L., & Farokhpay, M. (1985). Assessment of activity priorities and design preferences of elderly residents in public housing: A case study. *The Gerontologist, 25,* 251–257.

Nehrke, M. F., Morganti, J. B., Cohen, S. H., Hulicka, I. M., Whitbourne, S. K., Turner, R. R., & Cataldo, J. F. (1984). Differences in person-environment congruence between micro environments. *Canadian Journal on Aging, 3,* 117–132.

Nehrke, M. F., Turner, R. R., Cohen, S. H., Whitbourne, S. K., Morganti, J. B., & Hulicka, I. M. (1981). Toward a model of person-environment congruence: Development of the EPPIS. *Experimental Aging Research, 7,* 363–379.

Nelson, E. A., & Dannefer, D. (1992). Aged heterogeneity: Fact or fiction? The fate of diversity in gerontological research. *The Gerontologist, 32,* 17–23.

Nelson, G., & Earls, M. (1986). An action-oriented assessment of the housing and social support needs of long-term psychiatric clients. *Canadian Journal of Community Mental Health, 5,* 19–30.

Netten, A. (1989). The effect of design of residential homes in creating dependency among confused elderly residents: A study of elderly demented residents and their ability to find their way around homes for the elderly. *International Journal of Geriatric Psychiatry, 4,* 143–153.

Netten, A. (1991). Residential care and senile dementia: The effect of the physical and social environment of homes for elderly people on residents suffering from senile dementia (Doctoral dissertation, University of Kent, Canterbury, Kent, England, 1989). *Dissertation Abstracts International, 51,* 2841A.

Netten, A. (1991–92). A positive experience? Assessing the effect of the social environment on demented elderly residents of local authority homes. *Social Work and Social Sciences Review, 3,* 46–62.

Nirenberg, T. D. (1983). Relocation of institutionalized elderly. *Journal of Consulting and Clinical Psychology, 51,* 693–701.

Nyman, J. A. (1988). Improving the quality of nursing home outcomes: Are adequacy- or incentive-oriented policies more effective? *Medical Care, 26,* 1158–1171.

Nyman, J. A., & Geyer, C. R. (1989). Promoting the quality of life in nursing homes: Can regulation succeed? *Journal of Health Politics, Policy and Law, 14,* 797–816.

Oberle, K., Wry, J., & Paul, P. (1988). *The relationship between environment, anxiety and post-operative recovery.* Alberta, Canada: University of Alberta Hospitals, Department of Nursing.

Okun, M. A., Olding, R. W., & Cohn, C. M. G. (1990). A meta-analysis of subjective well-being interventions among elders. *Psychological Bulletin, 108,* 257–266.

Orchowsky, S. J. (1982). Person-environment interaction in nursing homes for the elderly (Doctoral dissertation, Virginia Commonwealth University, Charlottesville). *Dissertation Abstracts International, 43,* 1240B.

Osterburg, A. E. (1987). Evaluating design innovations in an extended care facility. In V. Regnier & J. Pynoos (Eds.), *Housing the aged: Design directives and policy considerations* (pp. 399–420). New York: Elsevier.

Palmer, M. A. (Ed.). (1981). *The architect's guide to facility programming.* Washington, DC: American Institute of Architects; New York: Architectural Record Books.

Parmelee, P. A., & Lawton, M. P. (1990). The design of special environments for the aged. In J. E. Birren & K. W. Schaie (Eds.), *Handbook of the psychology of aging* (3rd ed., pp. 464–488). San Diego, CA: Academic Press.

Patnaik, B., Lawton, M. P., Kleban, M. H., & Maxwell, R. (1974). Behavioral adaptation to the change in institutional residence. *The Gerontologist, 14,* 305–307.

Perkins, R. E., King, S. A., & Hollyman, J. A. (1989). Resettlement of old long-stay psychiatric patients: The use of the private sector. *British Journal of Psychiatry, 155,* 233–238.

Peters, H. J. M., & Boerma, L. H. (1983). Leefklimaat in verpleeguizen: Resultaat van een vooronderzoek. *Nederlands Tijdschrift Voor Sociale Gezondheidszorg, 61,* 278–279.

Philp, I., Mutch, W. J., Devaney, J., & Ogston, S. (1989). Can quality of life of old people in institutional care be measured? *Journal of Clinical Experimental Gerontology, 11,* 11–19.

Philp, I., Mutch, W. J., Ballinger, B. R., & Boyd, L. (1991). A comparison of care in private nursing homes, geriatric and psychogeriatric hospitals. *International Journal of Geriatric Psychiatry, 6,* 253–258.

Pruchno, R. A., & Resch, N. L. (1988). Intrainstitutional relocation: Mortality effects. *The Gerontologist, 28,* 311–317.

Pruchno, R. A., & Resch, N. L. (1989). The context of change: Disorientation following intrainstitutional relocation. *Journal of Applied Gerontology, 8,* 465–480.

Ramian, K. (1987). The resident oriented nursing home: A new dimension in the nursing home debate: Emphasis on living rather than nursing. *Danish Medical Bulletin,* Special Supplement Series No. 5, 89–93.

Rango, N. (1982). Nursing-home care in the United States. *New England Journal of Medicine, 307,* 883–889.

Regnier, V. (1987). Programming congregate housing: The preferences of upper income elderly. In V. Regnier & J. Pynoos (Eds.), *Housing the aged: Design directives and policy considerations* (pp. 207–226). New York: Elsevier.

Regnier, V., & Pynoos, J. (Eds.). (1987). *Housing the aged: Design directives and policy considerations.* New York: Elsevier.

Reich, J. W., & Zautra, A. J. (1990). Dispositional control beliefs and the consequences of a control-enhancing intervention. *Journal of Gerontology: Psychological Services, 45,* P46-P51.

Retsinas, J., & Garrity, P. (1985). Nursing home friendships. *The Gerontologist, 25,* 376–381.

Revicki, D. A., & May, H. J. (1989). Organizational characteristics, occupational stress, and mental health in nurses. *Behavioral Medicine, 15,* 30–36.

Rodin, J. (1983). Behavioral medicine: Beneficial effects of self-control training in aging. *International Review of Applied Psychology, 32,* 153–181.

Rodin, J. (1986). Aging and health: Effects of the sense of control. *Science, 233,* 1271–1276.

Rodin, J., Timko, C., & Harris, S. (1985). The construct of control: Biological and psychological correlates. In M. P. Lawton & G. L. Maddox (Eds.), *Annual review of gerontology and geriatrics* (Vol. 5, pp. 3–55). New York: Springer.

Rosow, I. (1967). *Social integration of the aged.* New York: Free Press.

Ross, J. K. (1977). *Old people, new lives: Community creation in a retirement residence.* Chicago: University of Chicago Press.

Rovins, G. (1990). Exploring the environmental effectiveness of normalization principles for older persons with developmental disabilities. *Adult Residential Care Journal, 4,* 37–49.

Rowe, J. W., & Kahn, R. L. (1987). Human aging: Usual and successful. *Science, 237,* 143–149.

Russell, R. V. (1990). Recreation and quality of life in old age: A causal analysis. *Journal of Applied Gerontology, 9,* 77–90.

Rutman, D. L., & Freedman, J. L. (1988). Anticipating relocation: Coping strategies and the meaning of home for older people. *Canadian Journal on Aging, 7,* 17–31.

Ryden, M. B. (1985). Environmental support for autonomy in the institutionalized elderly. *Research in Nursing and Health, 8,* 363–371.

Ryff, C. D., & Essex, M. J. (1991). Psychological well-being in adulthood and old age: Descriptive markers and explanatory processes. In K. W. Schaie (Ed.), *Annual review of geriatrics and gerontology* (Vol. 11, pp. 144–171). New York: Springer.

Schlenker, R. E., & Shaughnessy, P. W. (1984). Case mix, quality, and cost relationships in Colorado nursing homes. *Health Care Financing Review, 6,* 61–71.

Schlesinger, M., & Dorwart, R. (1984). Ownership and mental-health services: A reappraisal of the shift toward privately owned facilities. *New England Journal of Medicine, 311,* 959–965.

Schmidt, M. G. (1990). *Negotiating a good old age: Challenges of residential living in late life.* San Francisco: Jossey-Bass.

Schulz, R. (1976). Effects of control and predictability on the physical and psychological well-being of the institutionalized aged. *Journal of Personality and Social Psychology, 33,* 563–573.

Schulz, R., & Hanusa, B. H. (1980). Environmental influences on the effectiveness of control and competence-enhancing interventions. In L. C. Perlmutter & R. A. Monty (Eds.), *Choice and perceived control* (pp. 315–337). New York: Erlbaum.

Schulz, R., Heckhausen, J., & Locher, J. L. (1991). Adult development, control, and adaptive functioning. *Journal of Social Issues, 47*(4), 177–196.

Sechrest, L., West, S., Phillips, M., Redner, R., & Yeaton, W. (1979). Some neglected problems in evaluation research: Strength and integrity of treatments. In L. Sechrest, S. West, M. Phillips, R. Redner, & W. Yeaton (Eds.), *Evaluation studies review annual* (Vol. 4, pp. 15–35). Beverly Hills, CA: Sage.

Segal, S. P., & Aviram, U. (1978). *The mentally ill in community-based sheltered care: A study of community care and social integration.* New York: Wiley.

Sewell, F., & Lethaby, A. (1990). *Auckland homes for the elderly: Environmental quality and programmes.* Auckland, New Zealand: University of Auckland, Department of Community Health.

Shadish, W. R. (1989). Private-sector care for chronically mentally ill individuals: The more things change, the more they stay the same. *American Psychologist, 44,* 1142–1147.

Shadish, W. R., Orwin, R. G., Silber, B. G., & Bootzin, R. R. (1985). The subjective well-being of mental patients in nursing homes. *Evaluation and Program Planning, 8,* 239–250.

Sherman, S. R. (1974). Leisure activities in retirement housing. *Journal of Gerontology, 29,* 325–335.

Sherwood, S., Greer, D. S., Morris, J. N., & Mor, V. (1981). *An alternative to institutionalization: The Highland Heights experiment.* Cambridge, MA: Ballinger.

Sherwood, S., & Mor, V. (1980). Mental health institutions and the elderly. In J. E. Birren & R. B. Sloane (Eds.), *Handbook of mental health and aging* (pp. 854–884). Englewood Cliffs, NJ: Prentice-Hall.

Slivinske, L. R., & Fitch, V. L. (1987). The effect of control enhancing interventions on the well-being of elderly individuals living in retirement communities. *The Gerontologist, 27,* 176–181.

Smith, G. C. (1984). Perceptions of a geriatric health related facility by residents, staff, and observers: Validation of the Sheltered Care Environment Scale (Doctoral dissertation, University of Rochester, NY, 1983). *Dissertation Abstracts International, 45,* 910A.

Smith, G. C., & Whitbourne, S. K. (1990). Validity of the Sheltered Care Environment Scale. *Psychology and Aging, 5,* 228–235.

Smith, M., & Buckwalter, K. C. (1990). *Perceptions of the social climate during the process of change: The effects of relocation on a long term care center.* Cedar Rapids, IA: Abbe Center for Community Mental Health.

Smothers, B. (1987). The relationships among patient functional behaviors, patient characteristics, staff practices, and social climate in a psychiatric hospital (Doctoral dissertation, University of Maryland, College Park, 1986). *Dissertation Abstracts International, 48,* 1930B.

Smyer, M. A. (1989). Nursing homes as a setting for psychological practice: Public policy perspectives. *American Psychologist, 44,* 1307–1314.

Sommer, R. (1983). *Social design: Creating buildings with people in mind.* Englewood Cliffs, NJ: Prentice-Hall.

Spector, W. D., & Drugovich, M. L. (1989). Reforming nursing home quality regulation: Impact on cited deficiencies and nursing home outcomes. *Medical Care, 27,* 789–801.

Sperbeck, D. J., Whitbourne, S. K., & Nehrke, M. F. (1981). Determinants of person-environment congruence in institutionalized elderly men and women. *Experimental Aging Research, 7,* 381–392.

Spinner, B. J. H. (1986). Nursing home staff training in a modified aging microcosm module: Social climate and attitude change as perceived by residents and staff (Doctoral dissertation, University of Mississippi, 1985). *Dissertation Abstracts International, 46,* 3830A.

Stein, S., Linn, M. W., & Stein, E. M. (1987). Patients and staff assess social climate of different quality nursing homes. *Comprehensive Gerontology, 1,* 41–46.

Steinfeld, E. (1987). Adapting housing for older disabled people. In V. Regnier & J. Pynoos (Eds.), *Housing the aged: Design directives and policy considerations* (pp. 307–339). New York: Elsevier.

Stephens, M. A. P., Kinney, J. M., & McNeer, A. E. (1986). Accommodative housing: Social integration of residents with physical limitations. *The Gerontologist, 26,* 176–180.

Stirling, G., & Reid, D. W. (1992). The application of participatory control to facilitate patient well-being: An experimental study of nursing impact on geriatric patients. *Canadian Journal of Behavioural Science, 24,* 204–219.

Stones, M. J., Dornan, B., & Kozma, A. (1989). The prediction of mortality in elderly institution residents. *Journal of Gerontology, 44,* P72-P79.

Stones, M. J., & Kozma, A. (1989). Happiness and activities in later life: A propensity formulation. *Canadian Psychology, 30,* 526–537.

Strahan, G. (1987). Nursing home characteristics: Preliminary data from the 1985 National Nursing Home Survey. *Advanced data from vital and health statistics* (Report No. 131) (DHHS Publication No. PHS 87–1250). Hyattsville, MD: National Center for Health Statistics.

Strain, L. A. (1991). Use of health services in later life: The influence of health beliefs. *Journal of Gerontology: Social Sciences, 46,* S143-S150.

Svensson, T. (1984). *Aging and environment: Institutional aspects* (Linkoping Studies in Education Dissertations, No. 21). Linkoping, Sweden: Linkoping University.

Teitelman, J. L., & Priddy, J. M. (1988). From psychological theory to practice: Improving frail elders' quality of life through control enhancing interventions. *Journal of Applied Gerontology, 7,* 298–315.

Tellis-Nayak, V., & Tellis-Nayak, M. (1989). Quality of care and the burden of two cultures: When the world of the nurse's aide enters the world of the nursing home. *The Gerontologist, 29,* 307–313.

Tesch, S. A., Nehrke, M. F., & Whitbourne, S. K. (1989). Social relationships, psychosocial adaptation, and intrainstitutional relocation of elderly men. *The Gerontologist, 29,* 517–523.

Thompson, B., & Swisher, M. (1983). An assessment, using the Multiphasic Environmental Assessment Procedure (MEAP), of a rural life-care residential center for the elderly. *Journal of Housing for the Elderly, 1,* 41–56.

Timko, C., & Moos, R. H. (1989). Choice, control, and adaptation among elderly residents of sheltered care settings. *Journal of Applied Social Psychology, 19,* 636–655.

Timko, C., & Moos, R. H. (1990). Determinants of interpersonal support and self-direction in group residential facilities. *Journal of Gerontology: Social Sciences, 45,* S184-S192.

Timko, C., & Moos, R. (1991a). Assessing the quality of residential programs: Methods and applications. *Adult Residential Care Journal, 5,* 113–129.

Timko, C., & Moos, R. H. (1991b). A typology of social climates in group residential facilities for older people. *Journal of Gerontology: Social Sciences, 46,* S160-S169.

Tisdale, S. (1987). *Harvest moon: Portrait of a nursing home.* New York: Holt.

Townsend, P. (1962). *The last refuge.* London: Routledge and Paul.

Ullmann, S. G. (1981). Assessment of facility quality and its relationship to facility size in the long-term health care industry. *The Gerontologist, 21,* 91–97.

Ullmann, S. (1985). The impact of quality on cost in the provision of long-term care. *Inquiry, 22,* 293–302.

Ullmann, S. G. (1986). Chain ownership and long-term health care facility performance. *Journal of Applied Gerontology, 5,* 51–63.

Ullmann, S. G. (1987). Ownership, regulation, quality assessment, and performance in the long-term health care industry. *The Gerontologist, 27,* 233–239.

Vallerand, R. J., O'Connor, B. P., & Blais, M. R. (1989). Life satisfaction of elderly individuals in regular community housing, in low-cost community housing, and high and low self-determination nursing homes. *International Journal of Aging and Human Development, 28,* 277–283.

Veterans Administration. (1981). *1980 annual report.* Washington, DC: U.S. Government Printing Office.

Vladeck, B. C. (1980). *Unloving care: The nursing home tragedy.* New York: Basic Books.

Voeks, S. K., & Drinka, P. J. (1990). Participants' perception of a work therapy program in a nursing home. *Activities, Adaptation and Aging, 14,* 27–34.

Waters, J. E. (1980a). The social ecology of long-term care facilities for the aged: A case example. *Journal of Gerontological Nursing, 6,* 155–160.

Waters, J. E. (1980b). Systematic measures of ideal nursing homes: Planning and research prospects. *Aging and Leisure Living, 3,* 14–16.

Waxman, H. M., Carner, E. A., & Berkenstock, G. (1984). Job turnover and job satisfaction among nursing home aides. *The Gerontologist, 24,* 503–509.

Weihl, H. (1981). On the relationship between the size of residential institutions and the well-being of residents. *The Gerontologist, 21,* 247–250.

Weisbrod, B. A. (1989). Rewarding performance that is hard to measure: The private nonprofit sector. *Science, 244,* 541–546.

Weissert, W. G., Elston, J. M., Bolda, E. J., Cready, C. M., Zelman, W. N., Sloane, P. D., Kalsbeek, W. D., Mutran, E., Rice, T. H., & Koch, G. G. (1989). Models of adult day care: Findings from a national survey. *The Gerontologist, 29,* 640–649.

Wells, L. M., & Singer, C. (1988). Quality of life in institutions for the elderly: Maximizing well-being. *The Gerontologist, 28,* 266–269.

Wells, L. M., Singer, C., & Polgar, A. T. (1986). *To enhance quality of life in institutions: An empowerment model in long-term care: A partnership of residents, staffand families.* Toronto, Ontario, Canada: University of Toronto Press.

Wieland, D., Rubenstein, L. Z., Ouslander, J. G., & Martin, S. E. (1986). Organizing an academic nursing home: Impacts on institutionalized elderly. *Journal of the American Medical Association, 255,* 2622–2627.

Wiener, J. M., Hanley, R., Clark, R., & Van Nostrand, J. F. (1990). Measuring the activities of daily living: Comparisons across national surveys. *Journal of Gerontology: Social Sciences, 45,* S229-S237.

Wilkin, D., & Hughes, B. (1987). Residential care of elderly people: The consumers' views. *Aging and Society, 7,* 175–201.

Willcocks, D., Peace, S., & Kellaher, L. (1987). *Private lives in public places: A research-based critique of residential life in local authority old people's homes.* London: Tavistock Publications.

Wilson, J., & Kouzi, A. (1990). Quality of the residential environment in board-and-care homes for mentally and developmentally disabled persons. *Hospital and Community Psychiatry, 41,* 314–318.

Wingard, D. L., Jones, D. W., & Kaplan, R. M. (1987). Institutional care utilization by the elderly: A critical review. *The Gerontologist, 27,* 156–163.

Winkler, A., Fairnie, H., Gericevick, F., & Long, M. (1989). The impact of a resident dog on an institution for the elderly: Effects on perceptions and social interactions. *The Gerontologist, 29,* 216–223.

Winnett, R. L. (1989). Long-term care reconsidered: The role of the psychologist in the geriatric rehabilitation milieu. *Journal of Applied Gerontology, 8,* 53–68.

Wright, L. K. (1988). A reconceptualization of the "negative staff attitudes and poor care in nursing homes" assumption. *The Gerontologist, 28,* 813–820.

Author Index

Subject Index